Selected Titles in This Series

The Regulators of Beilinson and Borel

Volume 15

CRM MONOGRAPH SERIES

Centre de Recherches Mathématiques
Université de Montréal

The Regulators of Beilinson and Borel

José I. Burgos Gil

The Centre de Recherches Mathématiques (CRM) of the Université de Montréal was created in 1968 to promote research in pure and applied mathematics and related disciplines. Among its activities are special theme years, summer schools, workshops, postdoctoral programs, and publishing. The CRM is supported by the Université de Montréal, the Province of Québec (FCAR), and the Natural Sciences and Engineering Research Council of Canada. It is affiliated with the Institut des Sciences Mathématiques (ISM) of Montréal, whose constituent members are Concordia University, McGill University, the Université de Montréal, the Université du Québec à Montréal, and the Ecole Polytechnique. The CRM may be reached on the Web at www.crm.umontreal.ca.

American Mathematical Society
Providence, Rhode Island USA

Partially supported by grant DGICYT PB96-0234.

The production of this volume was supported in part by the Fonds pour la Formation de Chercheurs et l'Aide à la Recherche (Fonds FCAR) and the Natural Sciences and Engineering Research Council of Canada (NSERC).

2000 *Mathematics Subject Classification.* Primary 19F27; Secondary 14G10.

ABSTRACT. In this book we give a complete proof of the fact that Borel's regulator map is twice Beilinson's regulator map. The strategy of the proof follows the argument sketched in Beilinson's original paper and relies on very similar descriptions of the Chern–Weil morphisms and the van Est isomorphism.

The book also reviews some material from Algebraic Topology and Lie Group Theory needed in the comparison theorem.

Library of Congress Cataloging-in-Publication Data

Burgos Gil, José I. (José Ignacio), 1962–
The regulators of Beilinson and Borel / José I. Burgos Gil.
p. cm. — (CRM monograph series, ISSN 1065-8599 ; v. 15)
Includes bibliographical references and index.
ISBN 0-8218-2630-1 (alk. paper)
1. Regulators (Mathematics) I. Series.

QA247.B855 2001
512′.74—dc21 2001053829

∞ The paper used in this book is acid-free and falls within the guidelines
established to ensure permanence and durability.
This volume was submitted to the American Mathematical Society
in camera ready form by the Centre de Recherches Mathématiques.
Visit the AMS home page at URL: http://www.ams.org/

10 9 8 7 6 5 4 3 2 1 07 06 05 04 03 02

A Esther y Andrés,
María Pilar y Fausto,
Georgina, Fernando y F. F.,
Belén y Fausto.

Contents

Acknowledgments

I should like to express my gratitude to the organizers of the NATO ASI/CRM Summer School at Banff, "The Arithmetic and Geometry of Algebraic Cycles", B. Brent Gordon, James D. Lewis, Stefan Müller-Stach, Shuji Saito, and Noriko Yui, for their invitation to the Summer School, where the idea of writing this monograph took form, and for their kind support during the preparation of the manuscript. I am also indebted to Christophe Soulé and Vicente Navarro Aznar for their help and many useful discussions. My thanks to Boas Erez, Juan Carlos Naranjo and Herbert Gangl for their careful reading of the manuscript, their corrections and suggestions. Finally I would like to thank the referee for her or his competent and detailed comments and corrections.

CHAPTER 1

Introduction

The aim of this book is to give a complete proof of the fact that Borel's regulator map is twice Beilinson's regulator map (Theorem 10.9). The key ingredient in the proof is that the Chern–Weil morphism and the van Est isomorphism can be described explicitly in a very similar way (see Theorem 8.12 and Theorem 8.15).

Let us start recalling Dirichlet's regulator. Let k be a number field, let $\mathfrak{v}$ be its ring of integers and let $\mathfrak{v}^*$ be the group of units of $\mathfrak{v}$. Let r_1 (resp. $2r_2$) be the number of real (resp. complex) immersions of k. In his study of $\mathfrak{v}^*$, Dirichlet introduced a map

$$\rho\colon \mathfrak{v}^* \longrightarrow \mathbb{R}^{r_1+r_2}.$$

The image of this map is contained in a hyper-plane H. Moreover $\rho(\mathfrak{v}^*)$ is a lattice of H. That is

$$\rho\otimes\mathbb{R}\colon \mathfrak{v}^*\otimes\mathbb{R} \to H.$$

is an isomorphism. In particular the rank of $\mathfrak{v}^*$ is r_1+r_2-1. Let $R_{\mathrm{D}} = \operatorname{Vol}\bigl(H/\rho(\mathfrak{v}^*)\bigr)$ be the covolume of this lattice. This number is called Dirichlet's regulator. The most interesting fact about this regulator is the class number formula:

$$R_{\mathrm{D}} = -\frac{w}{h}\lim_{s\to 0}\zeta_k(s)s^{-(r_1+r_2-1)}, \tag{1.1}$$

where ζ_k is Dedekind's zeta function of the field k, w is the number of roots of unity and h is the class number. Since Dedekind's zeta function is defined using local data at the primes of $\mathfrak{v}$, this formula can be seen as a highly non trivial local to global principle.

Recall that $\mathfrak{v}^*$ is the K-theory group $K_1(\mathfrak{v})$. In order to generalize formula (1.1) to higher K-theory, Borel [**5**] has introduced, for all $p\geq 2$, morphisms

$$r'_{\mathrm{Bo}}\colon K_{2p-1}(\mathfrak{v}) \longrightarrow V_p,$$

where V_p is a real vector space of dimension

$$\dim_{\mathbb{R}} V_p = d_p = \begin{cases} r_1+r_2, & \text{if } p \text{ is odd},\\ r_2, & \text{if } p \text{ is even}.\end{cases}$$

These morphisms will be called Borel's regulator maps. Moreover Borel has proved that $r'_{\mathrm{Bo}}\bigl(K_{2p-1}(\mathfrak{v})\bigr)$ is a lattice of V_p. As a consequence he obtains that the rank of the group $K_{2p-1}(\mathfrak{v})$ is d_p.

Lichtenbaum in [**42**] asked that, if one chooses a natural lattice L' in V_p, and defines

$$R'_{\mathrm{Bo},p} = \mathrm{CoVol}\big(r'_{\mathrm{Bo}}\big(K_{2p-1}(\mathfrak{v})\big), L'\big),$$

whether it is true that

$$(1.2) \qquad R'_{\mathrm{Bo},p} = \pm \frac{\sharp K_{2p-2}(\mathfrak{v})}{\sharp K_{2p-1}(\mathfrak{v})_{\mathrm{tor}}} \lim_{s\to -p+1} \zeta_k(s)(s+p-1)^{-d_p}.$$

Lichtenbaum gave a concrete choice of lattice L', but pointed out that, due to the lack of examples at that time, it might be necessary to adjust the formula by some power of π and some rational number.

In [**8**] Borel proved that

$$R'_{\mathrm{Bo},p} \sim \pi^{-d_p} \lim_{s\to -p+1} \zeta_k(s)(s+p-1)^{-d_p},$$

where $a \sim b$ means that there exists an element $q \in \mathbb{Q}^*$ such that $qa = b$. The number $R'_{\mathrm{Bo},p}$ is called Borel's regulator.

REMARK 1.1. The subindexes used here do not agree with the convention used in [**8**]. In particular the regulator $R'_{\mathrm{Bo},p}$ is R_{p-1} in the notation of [**8**].

The factor π^{-d_p} means that the original choice of lattice was not the best one. Moreover, the original definition of Borel does not factorize through the K-theory of the field $\mathbb{C}$. For these reasons it is convenient to renormalize Borel's regulator map. This renormalized regulator usually appears in the literature instead of the original definition. We will denote the renormalized Borel regulator map as r_{Bo}.

The relationship between values of zeta functions, or more generally, L-functions and regulators is a very active field with many open conjectures. Beilinson has generalized the definition of regulators and stated very general conjectures relating values of L functions and regulators associated with algebraic motives.

One can see Borel's theorem as the Beilinson conjecture in the case of number fields as follows. Let us write $X = \operatorname{Spec} \mathfrak{v}$. Then, for $p \geq 0$,

$$K_{2p-1}(\mathfrak{v}) \otimes \mathbb{Q} = H^1_{\mathcal{A}}(X, \mathbb{Q}(p))$$

where the right hand side is called absolute cohomology. In general, rational absolute cohomology is a graded piece for a certain filtration of rational K-theory. But in the case of number fields there is only one non zero piece.

The Chern character for higher K-theory induces a morphism

$$r_{\mathrm{Be}} \colon H^1_{\mathcal{A}}(X, \mathbb{Q}(p)) \to H^1_{\mathcal{D}}(X_{\mathbb{R}}, \mathbb{R}(p)),$$

such that $r_{\mathrm{Be}} \otimes \mathbb{R}$ is an isomorphism. The morphism r_{Be} is called Beilinson's regulator map. The Deligne–Beilinson cohomology group $H^1_{\mathcal{D}}(X_{\mathbb{R}}, \mathbb{R}(p))$, has a natural rational structure and Beilinson's regulator is the determinant of $\mathrm{Im}(r_{\mathrm{Be}})$ with respect to this rational structure. We will denote it as $R_{\mathrm{Be},p}$.

Observe that, in this setting, Beilinson's regulator is defined only up to a rational number. For number fields, Beilinson's conjectures state that

$$R_{\mathrm{Be},p} \sim \lim_{s \to -p+1} \zeta_k(s)(s+p-1)^{-d_p}.$$

In order to see Borel's theorem as a particular case of Beilinson's conjectures we have to compare the two regulators.

To this end, we consider Beilinson's regulator map as a morphism

$$r_{\mathrm{Be}} \colon K_{2p-1}(\mathcal{O}) \to H^1_{\mathcal{D}}\big(X_{\mathbb{R}}, \mathbb{R}(p)\big).$$

Moreover, $H^1_{\mathcal{D}}\big(X_{\mathbb{R}}, \mathbb{R}(p)\big)$ contains a natural lattice L and we can write

$$R_{\mathrm{Be},p} = \operatorname{CoVol}\big(r_{\mathrm{Be}}\big(K_{2p-1}(\mathfrak{v})\big), L\big).$$

This is a well defined real number.

In [**2**], Beilinson claims that $r_{\mathrm{Bo}} = r_{\mathrm{Be}}$ and gave the sketch of a proof of this fact. Rapoport [**55**] completed most of Beilinson's proof and showed that r_{Bo} and r_{Be} agree up to a rational number. In [**25**], Dupont, Hain and Zucker gave a completely different strategy to try to compare the regulators. Moreover they conjectured that the precise comparison is $r_{\mathrm{Bo}} = 2r_{\mathrm{Be}}$. In this book we will use Beilinson's original argument to show that indeed $r_{\mathrm{Bo}} = 2r_{\mathrm{Be}}$. Except for the precise comparison, there is very little original in this book: the original argument is due to Beilinson and we follow Rapoport's paper in several points.

One of the difficulties a beginner may have in studying this topic is the maze of cohomology theories used and the different results from algebraic topology and Lie group theory needed. For the convenience of the reader we have included an introduction to different topics, such as simplicial techniques, Hopf algebras, Chern–Weil theory, Lie algebra cohomology and continuous group cohomology. A complete treatment of each of these areas would merit a book on his own and there are many of them available. Therefore in these introductions only the results directly related with the definition and comparison of the regulators are stated and most of them without proof. With this idea, the book can be divided in two parts. The first one, from Chapter 2 to Chapter 6, is a collection of classical results. The main purpose of this part is to aid understanding of both regulator maps and to fix the notations. So a reader may skip some of these chapters and refer to them if needed. The second part, from Chapter 7 to Chapter 10, is the heart of the work. It contains the definition of the regulator maps and the specific tools needed for the comparison.

Let us give a more detailed account of the contents of each chapter. In Chapter 2 we recall the definition and some properties of simplicial and cosimplicial objects. We also give the definition of sheaves and principal bundles over simplicial spaces and we recall Dupont's definition of the de Rham algebra of a simplicial differentiable manifold. Chapter 3 is devoted to H-spaces and Hopf algebras. The main results are the structure theorems of Hopf algebras and the relationship between the homotopy and the primitive part of the homology of an H-space. In Chapter 4 we compute the singular cohomology of the general linear group and of its classifying space. The cohomology of these spaces is related by the suspension, or its inverse, the transgression. This map is one of the ingredients of the comparison between the regulators. We also recall Bott's Periodicity Theorem that characterizes the stable homotopy of the classical groups. In Chapter 5 we review the de Rham cohomology of Lie groups and its relationship with Lie algebra cohomology. We also recall the definition of the Weil algebra and the Chern–Weil theory of characteristic classes from the de Rham point of view. We show that the suspension can be computed using the Weil algebra. We also give explicit representatives of the generators of the cohomology of the Lie algebra $\mathfrak{u}_n$. We end the chapter recalling the definition of relative Lie algebra cohomology. In Chapter 6 we give an introduction to continuous group cohomology. We also recall the construction of the the van Est isomorphism relating continuous group cohomology and relative Lie algebra cohomology. We will see how the van Est isomorphism allows us to compute the continuous cohomology of the classical groups. Both regulators are determined by classes in continuous group cohomology. To compare the regulators we will compare these classes. Chapter 7 is devoted to the theory of small cosimplicial algebras and small differential graded algebras. This theory was introduced by Beilinson to compare the regulators. In this chapter we will follow Rapoport's paper closely. In Chapter 8 we give a description of the sheaf of differential forms as the sheaf of functions on certain simplicial scheme modulo a sheaf of ideals. This description is a generalization of the fact that the sheaf of 1-forms can be written as the ideal of the diagonal modulo its square. This description is the main ingredient for the comparison and it is implicit in Guichardet's description of the van Est isomorphism [**35**]. In Chapter 9 we recall the definition of algebraic K-theory and the definition of Borel's regulator. We also discuss the renormalization of Borel's regulator and we give an explicit representative of the cohomology class of Borel's regulator in Lie algebra cohomology. Finally in Chapter 10 we recall the definition of Beilinson's regulator and we prove the comparison theorem.

CHAPTER 2

Simplicial and Cosimplicial Objects

2.1. Basic Definitions and Examples

In this section we will recall the definition and properties of simplicial and cosimplicial objects and give some examples. The main purpose is to fix the notation. For more details, the reader is referred to [**12, 28, 31, 44**].

Let Δ be the category whose objects are the ordinal numbers

$$[n] = \{0, \dots, n\},$$

and whose morphisms are the increasing maps between them. The morphisms of the category Δ are generated by the morphisms

$$\begin{aligned} \delta^i &: [n-1] \to [n], && \text{for } n \geq 1,\ i = 0, \dots, n, \\ \sigma^i &: [n+1] \to [n], && \text{for } n \geq 0,\ i = 0, \dots, n, \end{aligned}$$

where

$$\delta^i(k) = \begin{cases} k, & \text{if } k < i, \\ k+1, & \text{if } k \geq i, \end{cases} \qquad \sigma^i(k) = \begin{cases} k, & \text{if } k \leq i, \\ k-1, & \text{if } k > i. \end{cases}$$

The morphism δ^i are called *faces* and the morphisms σ^i are called *degeneracies*. These morphisms satisfy the following commutation rules

$$\begin{aligned} \delta^j \delta^i &= \delta^i \delta^{j-1}, && \text{for } i < j, \\ \sigma^j \sigma^i &= \sigma^i \sigma^{j+1}, && \text{for } i \leq j, \\ \sigma^j \delta^i &= \delta^i \sigma^{j-1}, && \text{for } i < j, \\ \sigma^j \delta^i &= \mathrm{Id}, && \text{for } i = j, j+1, \\ \sigma^j \delta^i &= \delta^{i-1} \sigma^j, && \text{for } i > j+1. \end{aligned} \tag{2.1}$$

We will denote the opposite category of Δ by Δ^{op}: that is, the category with the same objects but reversed arrows.

DEFINITION 2.1. Let $\mathcal{C}$ be a category. A *simplicial object* of $\mathcal{C}$ is a functor $\Delta^{\mathrm{op}} \to \mathcal{C}$. A *cosimplicial object* of $\mathcal{C}$ is a functor $\Delta \to \mathcal{C}$. The category of simplicial objects of $\mathcal{C}$ will be denoted by $\mathcal{S}(\mathcal{C})$ and the category of cosimplicial objects by $\mathcal{CS}(\mathcal{C})$.

In other words, a simplicial object of $\mathcal{C}$ is a family of objects of $\mathcal{C}$, $\{X_n\}_{n\geq 0}$, together with morphisms

$$\begin{aligned} \delta_i &\colon X_n \to X_{n-1}, && \text{for } n \geq 1,\ i = 0, \dots, n, \\ \sigma_i &\colon X_n \to X_{n+1}, && \text{for } n \geq 0,\ i = 0, \dots, n, \end{aligned}$$

satisfying the commutation rules

$$\begin{aligned} \delta_i\delta_j &= \delta_{j-1}\delta_i, && \text{for } i < j, \\ \sigma_i\sigma_j &= \sigma_{j+1}\sigma_i, && \text{for } i \leq j, \\ \delta_i\sigma_j &= \sigma_{j-1}\delta_i, && \text{for } i < j, \\ \delta_i\sigma_j &= \mathrm{Id}, && \text{for } i = j, j+1, \\ \delta_i\sigma_j &= \sigma_j\delta_{i-1}, && \text{for } i > j+1. \end{aligned} \tag{2.2}$$

The morphisms δ_i are also called faces and the morphisms σ_i degeneracies. Sometimes, it will be usefull to use the functorial notation. That is, if

$$\tau\colon [n] \to [m]$$

is an increasing map, we denote by

$$X(\tau)\colon X_m \to X_n$$

the corresponding morphism. In particular $\delta_i = X(\delta^i)$ and $\sigma_i = X(\sigma^i)$.

Analogously, a cosimplicial object of $\mathcal{C}$ is a family of objects of $\mathcal{C}$, $\{X^n\}_{n\geq 0}$, together with morphisms

$$\begin{aligned} \delta^i &\colon X^n \to X^{n+1}, && \text{for } n \geq 0,\ i = 0, \dots, n+1, \\ \sigma^i &\colon X^n \to X^{n-1}, && \text{for } n \geq 1,\ i = 0, \dots, n-1, \end{aligned}$$

satisfying the commutation rules (2.1). Observe that we use the convention that simplicial objects are indexed using subscripts and cosimplicial objects are indexed by superscripts.

EXAMPLE 2.2. The *geometric simplex*, denoted $\underline{\Delta}^{\cdot}$, is the cosimplicial topological space defined by

$$\underline{\Delta}^n = \{(t_0, \dots, t_n) \in \mathbb{R}^{n+1} \mid t_0 + \dots + t_n = 1, t_i \geq 0\},$$

with faces and degeneracies given by

$$\begin{aligned} \delta^i(t_0, \dots, t_n) &= (t_0, \dots, t_{i-1}, 0, t_i, \dots, t_n), \\ \sigma^i(t_0, \dots, t_n) &= (t_0, \dots, t_{i-1}, t_i + t_{i+1}, t_{i+2}, \dots, t_n). \end{aligned} \tag{2.3}$$

EXAMPLE 2.3. Let X be a topological space. Then *the simplicial set of singular simplexes* of X is given by

$$S_n(X) = \mathrm{Hom}_{\mathbf{Top}}(\underline{\Delta}^n, X),$$

where **Top** denotes the category of topological spaces. If $f \in S_n(X)$ then

$$\begin{aligned} \delta_i f &= f \circ \delta^i, \\ \sigma_i f &= f \circ \sigma^i. \end{aligned}$$

$S_.$ is a functor between **Top** and the category of simplicial sets, $\mathcal{S}(\mathbf{Set})$. This functor will be called the singular functor. If G is an abelian group we will denote by $S_n(X,G)$ the free G module generated by $S_n(X)$ and by

$$S^n(X,G) = \operatorname{Hom}_{\mathbf{Ab}}(S_n(X,\mathbb{Z}),G).$$

Then $S_.(X,G)$ is the simplicial abelian group of singular chains and $S^{\cdot}(X,G)$ is the cosimplicial abelian group of singular cochains.

EXAMPLE 2.4. Let $\Delta[k]$ be the simplicial set such that $\Delta[k]_n$ consists of all the increasing maps from $[n]$ to $[k]$. In other words, $\Delta[k]_n$ is the set of all sequences $(j_0,\dots,j_n)$, with $0 \le j_0 \le \cdots \le j_n \le k$. The faces and degeneracies are given by

$$\begin{aligned} \delta_i(j_0,\dots,j_n) &= (j_0,\dots,\widehat{j_i},\dots,j_n), && \text{for } i=0,\dots,n,\\ \sigma_i(j_0,\dots,j_n) &= (j_1,\dots,j_i,j_i,\dots,j_n), && \text{for } i=0,\dots,n, \end{aligned}$$

where the symbol $\widehat{j_i}$, means that the element j_i is omitted. For instance $\Delta[0]$ is the simplicial set with one element in each degree. The only element that is non degenerate is the element in degree 0. The simplicial set $\Delta[1]$ has three non degenerate elements: one in degree one and two in degree zero.

The increasing maps between $[k]$ and $[k']$ induce maps between $\Delta[k]$ and $\Delta[k']$. For instance the maps $\delta^0,\delta^1\colon [0] \to [1]$ induce maps $\delta^0,\delta^1\colon \Delta[0] \to \Delta[1]$. Thus $\Delta[\cdot]$ is a cosimplicial simplicial set. The simplicial set $\Delta[k]$ plays the role of the geometric k-dimensional simplex. In this analogy, the above maps from $\Delta[0]$ to $\Delta[1]$ correspond to the inclusions of a point as each of the vertexes of the unit interval.

EXAMPLE 2.5. Let X be a topological space. Let $E_.X$ be the simplicial topological space defined by

$$\begin{aligned} E_nX &= \overbrace{X\times\cdots\times X}^{n+1}\\ \delta_i(x_0,\dots,x_n) &= (x_0,\dots,\widehat{x_i},\dots,x_n), && \text{for } i=0,\dots,n,\\ \sigma_i(x_0,\dots,x_n) &= (x_1,\dots,x_i,x_i,\dots,x_n), && \text{for } i=0,\dots,n. \end{aligned}$$

Observe that we can define in an analogous way $E_.X$ for X a differentiable manifold, a scheme over a base scheme, or more generally in any category with finite products.

If $\mathcal{C}$ is a category with products then the categories $\mathcal{S}(\mathcal{C})$ and $\mathcal{CS}(\mathcal{C})$ also have products. For instance, the product of two simplicial objects X and Y is given explicitly by

$$(X\times Y)_n = X_n \times Y_n,$$

with faces and degeneracies defined componentwise.

We will be also interested in the following construction. Let $\mathcal{C}$ be a category that admits coproducts. If X is an object of $\mathcal{S}(\mathcal{C})$ and K is a simplicial set, we may define an object of $\mathcal{S}(\mathcal{C})$, $X \times K$ by

$$(X \times K)_n = X_n \times K_n = \coprod_{p \in K_n} X_n,$$

where the faces and degeneracies are given componentwise. In particular this applies when X is an object of $\mathcal{C}$. Then we denote also by X the constant simplicial object, with $X_n = X$ and all faces and degeneracies equal to the identity. In this case

$$(X \times K)_n = \coprod_{p \in K_n} X.$$

Let X and Y be two simplicial objects of $\mathcal{C}$. Let $f, g : X \to Y$ be two simplicial morphisms. A simplicial homotopy between f and g is a simplicial morphism

$$H \colon X \times \Delta[1] \to Y,$$

such that $H \circ \mathrm{Id} \times \delta^0 = f$ and $H \circ \mathrm{Id} \times \delta^1 = g$.

REMARK 2.6. The homotopy relation is not, in general, an equivalence relation (see [**31**, I.6]). For instance, in the case of simplicial sets we need the condition of Y being fibrant.

PROPOSITION 2.7. *Let X be a topological space and let $e \in X$ be a point. Then the identity map* $\mathrm{Id} \colon E.X \to E.X$ *is homotopically equivalent to the constant map that sends $E_n X$ to the point* $(e, \dots, e)$.

SKETCH OF PROOF. We have to construct a morphism of simplicial topological spaces $H \colon E.X \times \Delta[1] \to E.X$ such that $H \circ (\mathrm{Id} \times \delta^0) = \mathrm{Id}$ and $H \circ (\mathrm{Id} \times \delta^1)$ is the constant map e. This morphism H is:

$$H\big((x_0, \dots, x_n), (i_0, \dots, i_n)\big) = \big(f_{i_0}(x_0), \dots, f_{i_n}(x_n)\big),$$

where $f_0(x) = x$ and $f_1(x) = e$. □

2.2. Simplicial Abelian Groups

Let $\mathbf{A}$ be an abelian category. Let us denote by $\mathcal{C}^+(\mathbf{A})$ the category of non negatively graded cochain complexes of $\mathbf{A}$. By the Dold–Kan correspondence (cf. [**31**, III.2]) there is an equivalence of categories between $\mathcal{C}^+(\mathbf{A})$ and $\mathcal{CS}(\mathbf{A})$. Analogously, there is also an equivalence between the category $\mathcal{S}(A)$ and the category of non negatively graded chain complexes, $\mathcal{C}_+(A)$. Let us recall this theory.

DEFINITION 2.8. Let X be an object of $\mathcal{CS}(\mathbf{A})$. Then $(\mathcal{C}X, d)$ is the object of $\mathcal{C}^+(\mathbf{A})$ given by

$$\mathcal{C}X^n = X^n$$

$$dx = \sum_{i=0}^{n+1} (-1)^i \delta^i x, \quad \text{for } x \in \mathcal{C}X^n.$$

The *normalization* of X, denoted by $\mathcal{N}X$ is the subcomplex of $\mathcal{C}X$ defined by

$$\mathcal{N}X^n = \bigcap_{i=0}^{n-1} \operatorname{Ker} \sigma^i.$$

Let us write

$$\mathcal{D}X^n = \sum_{i=0}^{n-1} \operatorname{Im} \delta^i.$$

By the commutation rules (2.1), it is clear that $\mathcal{D}X = \bigoplus \mathcal{D}X^n$, is a subcomplex of $\mathcal{C}X^n$. For a proof of the following result see, for instance, [**31**, Chap. III].

PROPOSITION 2.9. *Let $X \in \operatorname{Ob}(\mathcal{CS}(\mathbf{A}))$. Let $i \colon \mathcal{N}X \to \mathcal{C}X$ be the inclusion, and let $p \colon \mathcal{C}X \to \mathcal{C}X/\mathcal{D}X$ be the projection. Then the composition*

$$\mathcal{N}X \xrightarrow{i} \mathcal{C}X \xrightarrow{p} \mathcal{C}X/\mathcal{D}X$$

is an isomorphism. Moreover the composition $i \circ (p \circ i)^{-1} \circ p$ is homotopically equivalent to the identity of $\mathcal{C}X$. In consequence $\mathcal{C}X$ and $\mathcal{N}X$ are homotopically equivalent complexes.

COROLLARY 2.10. *Let X be a cosimplicial abelian group. Then there is a direct sum decomposition*

$$X^p = \mathcal{N}X^p \oplus \mathcal{D}X^p.$$

EXAMPLE 2.11. Let X be a topological space and let G be an abelian group. We will write

$$C^*(X,G) = \mathcal{C}S^*(X,G).$$

This is the complex of singular G-cochains on X. The singular cohomology groups of X, with coefficients in G, are the cohomology groups of the complex $C^*(X,G)$. By Proposition 2.9 the cohomology groups of X are also the cohomology groups of the complex $\mathcal{N}S^*(X,G)$.

EXAMPLE 2.12. We can also use the normalization functor to define the singular cohomology of a simplicial topological space. Let $X.$ be a simplicial topological space. Then the complexes $C^*(X_n,G)$ form a cosimplicial complex. The normalization $\mathcal{N}C^*(X.,G)$ can be turned into a double complex, and the singular cohomology groups of $X.$ are the cohomology groups of the simple complex associated to this double complex.

Now let $Y \in \operatorname{Ob}(\mathcal{C}^+(\mathbf{A}))$. We want to construct an object of $\mathcal{CS}(\mathbf{A})$, $\mathcal{K}Y$, such that $\mathcal{NK}Y$ is naturally isomorphic to Y. The basic idea behind the construction is that we have to add enough degenerate elements in order to be able to define all faces and degeneracies.

DEFINITION 2.13. The *Dold–Kan functor* is the functor that associates, to each complex (Y, d), the cosimplicial object $\mathcal{K}Y$. This cosimplicial object has components

$$(\mathcal{K}Y)^n = \bigoplus_{f\colon [n]\to[p]} Y_f^p,$$

where the sum runs over all surjective increasing maps f, and $Y_f^p = Y^p$. The structure morphisms of $\mathcal{K}Y$ are constructed as follows. Let $u\colon [n] \to [m]$ be an increasing map. Then the morphism

$$\mathcal{K}Y(u)\colon (\mathcal{K}Y)^n \to (\mathcal{K}Y)^m$$

can be decomposed in components

$$\mathcal{K}Y(u)_{f,g}\colon Y_f^p \to Y_g^q,$$

for all pair of surjective increasing morphism $f\colon [n] \to [p]$ and $g\colon [m] \to [q]$. Then we write $\mathcal{K}Y(u)_{f,g} = \mathrm{Id}$ if $p = q$ and there exists a commutative diagram

$$\begin{array}{ccc} [n] & \xrightarrow{u} & [m] \\ {\scriptstyle f}\downarrow & & \downarrow{\scriptstyle g} \\ [p] & \xrightarrow{j} & [q]. \end{array}$$

with j the identity. We write $\mathcal{K}Y(u)_{f,g} = d$ if $[q] = [p+1]$ and there exists a diagram as above with $j = \delta^0$. Finally we write $\mathcal{K}Y(u)_{f,g} = 0$ in all other cases.

For a proof of the following result see for instance [**31**, III.2]

THEOREM 2.14 (Dold–Kan correspondence). *The functors $\mathcal{N}$ and $\mathcal{K}$ establish an equivalence of categories between $\mathcal{C}^+(\mathbf{A})$ and $\mathcal{CS}(\mathbf{A})$.*

2.3. The Geometric Realization

The main link between simplicial sets and topological spaces is the geometric realization. This functor, together with the singular functor of Example 2.3, establishes an equivalence between the homotopy categories of simplicial sets and of topological spaces. Thus, up to homotopy, both categories are equivalent. The geometric realization functor can be extended to the case of simplicial topological spaces.

DEFINITION 2.15. Let $X.$ be a simplicial topological space. The *geometric realization* is the topological space

$$|X.| = \coprod_{n\geq 0} X_n \times \underline{\Delta}^n \Big/ \sim,$$

where $\sim$ is the equivalence relation generated by

$$(\sigma_i(x), y) \sim \big(x, \sigma^i(y)\big) \quad \text{and} \quad (\delta_i(x), y) \sim \big(x, \delta^i(y)\big).$$

2.4. Sheaves on Simplicial Topological Spaces

Let us discuss sheaf cohomology for simplicial topological spaces. For more details the reader is referred to [**20**, §5] and [**30**].

DEFINITION 2.16. Let $X.$ be a simplicial topological space. A *sheaf of abelian groups* on $X.$ is the data of a sheaf $\mathcal{F}^n$ over each X_n together with morphisms

$$\delta^i \colon \mathcal{F}^{n-1} \to (\delta_i)_*\mathcal{F}^n \quad \text{and} \quad \sigma^i \colon \mathcal{F}^{n+1} \to (\sigma_i)_*\mathcal{F}^n$$

satisfying the commutation rules (2.1).

A morphism between two sheaves $\mathcal{F}^{\cdot}$ and $\mathcal{G}^{\cdot}$ is a family of morphisms of sheaves $\{f^n\}_{n\geq 0}$ commuting with the faces and degeneracies.

For instance, if $X.$ is a simplicial scheme, the family of structural sheaves $\{\mathcal{O}_{X_n}\}_{n\geq 0}$ is a sheaf on $X.$. The category of sheaves of abelian groups on a simplicial topological space is an abelian category.

DEFINITION 2.17. Let $X.$ be a simplicial topological space. Let $\mathcal{F}^{\cdot}$ be a sheaf of abelian groups on $X.$. Then the *group of global sections of* $\mathcal{F}$, denoted by $\Gamma(\mathcal{F})$, is the group

$$\Gamma(\mathcal{F}) = \{s \in \Gamma(X_0, \mathcal{F}^0) \mid \delta^0 s = \delta^1 s\}.$$

The *cohomology groups* of $\mathcal{F}^{\cdot}$, denoted by $H^*(X., \mathcal{F}^{\cdot})$ are the right derived functors of the functor Γ.

The sheaf cohomology groups can be computed using resolutions in the following way. For each sheaf $\mathcal{F}^n$, let $i^n \colon \mathcal{F}^n \to \mathcal{A}^{n,*}$ be a resolution by acyclic sheaves, such that $\mathcal{A}^{\cdot,*}$ is a complex of sheaves on $X.$ and the morphism $i = \{i^n\}$ is a morphism of sheaves on $X.$. For instance we can use the canonical flasque resolution of each sheaf $\mathcal{F}^n$. Then the cohomology groups of $\mathcal{F}^{\cdot}$ are the cohomology groups of the simple complex associated to the complex of complexes $\mathcal{N}\Gamma(X., \mathcal{A}^{\cdot,*})$:

$$\mathcal{N}\mathcal{A} = s\big(\mathcal{N}\Gamma(X., \mathcal{A}^{\cdot,*})\big).$$

Or, equivalently, to the cohomology of the complex

$$\mathcal{C}\mathcal{A} = s\big(\mathcal{C}\Gamma(X., \mathcal{A}^{\cdot,*})\big).$$

In the complex $\mathcal{C}\mathcal{A}$ we can introduce a filtration associated with the simplicial degree

$$F^p\mathcal{C}\mathcal{A} = \bigoplus_{\substack{n\geq p\\ m}} \Gamma(X_n, \mathcal{A}^{n,m}).$$

This filtration determines an spectral sequence (cf. [**20**, §5] and [**30**]).

PROPOSITION 2.18. *Let $X.$ be a simplicial topological space, and let $\mathcal{F}^{\cdot}$ be a sheaf over $X.$. Then there is a first quadrant spectral sequence*

$$E_1^{p,q} = H^q(X_p, \mathcal{F}^p) \implies H^{p+q}(X., \mathcal{F}^{\cdot}).$$

This section can be easily generalized to cover the case of complexes of sheaves on simplicial topological spaces.

2.5. Principal Bundles on Simplicial Manifolds

Let G be a Lie group. Recall that a principal G-bundle is a four-tuple (r, E, π, B), where E and B are differentiable manifolds, $\pi\colon E \to B$ is a morphism of differentiable manifolds and $r\colon E \times G \to E$ is a differentiable right action of G on E, such that there exists an open covering $\mathcal{U}$ of B and, for every open subset $U \in \mathcal{U}$, there is an isomorphism of differentiable manifolds

$$\phi_U : U \times G \to \pi^{-1}(U)$$

satisfying

(1) $\pi\big(\phi_U(x, g)\big) = x$.
(2) $\phi_U(x, gs) = r(\phi_U(x, g), s)$.

The manifold E is called the total space and B is called the base space. Usually we will denote a principal G bundle by its total space E and the right action r will be denoted by $r(p, s) = ps$.

Observe that the definition of a principal G-bundle implies that G acts freely on E and transitively on the fibres of π.

Let (r, E, π, B) be a principal G-bundle and let $f\colon \widehat{B} \to B$ be a morphism of differentiable manifolds; then we can define in an obvious way a principal G-bundle f^*E.

A morphism of principal G bundles,

$$f\colon (\hat{r}, \widehat{E}, \hat{\pi}, \widehat{B}) \to (r, E, \pi, B),$$

is a commutative diagram

$$\begin{array}{ccc} \widehat{E} & \xrightarrow{f_E} & E \\ {\scriptstyle\hat{\pi}}\downarrow & & \downarrow{\scriptstyle\pi} \\ \widehat{B} & \xrightarrow{f_B} & B \end{array} \tag{2.4}$$

such that $f_E(pg) = f_E(p)g$ for all $p \in \widehat{E}$ and $g \in G$. Clearly, for a fixed f_B, to give a morphism of principal G bundles as above is equivalent to give an isomorphism $\widehat{E} \cong f_B^*E$. We will denote this isomorphism also by f_E.

DEFINITION 2.19. A *simplicial principal G bundle* is a four-tuple $(r, E., \pi, B.)$, where $E.$ and $B.$ are simplicial differentiable manifolds, $\pi\colon E. \to B.$ is a morphism of simplicial manifolds and r is a right action of G on $E.$ such that, for all n, (r, E_n, π, B_n) is a principal G bundle and all faces and degeneracies are morphisms of principal G-bundles.

PROPOSITION 2.20. *Let G be a Lie group and let $B.$ be a simplicial differentiable manifold. Then there is an equivalence of categories between the category of simplicial principal G-bundles and the category of pairs (E, α), where E is a principal G-bundle over B_0 and*

$$\alpha : \delta_0^* E \to \delta_1^* E$$

is an isomorphism of principal G-bundles over B_1.

PROOF. Let us exhibit functors between the two categories. Let $E.$ be a principal G-bundle over $B.$. There are isomorphisms of principal G-bundles over B_1

$$E(\delta^0)\colon E_1 \to \delta_0^* E_0,$$
$$E(\delta^1)\colon E_1 \to \delta_1^* E_0.$$

The functor in one direction sends $E.$ to the pair $(E_0, E(\delta^1) \circ E(\delta^0)^{-1})$.

Let us construct the functor in the other direction. Let (E, α) be a pair as in the proposition. Then we write

$$E_n = \left((\delta_0)^n\right)^* E.$$

Let

$$\tau\colon [n] \to [m]$$

be an increasing map. We have to construct a principal G-bundle morphism $E(\tau)\colon E_m \to E_n$. Or, equivalently, an isomorphism, also denoted $E(\tau)$

$$E(\tau)\colon B\left((\delta^0)^m\right)^* E \to B(\tau \circ (\delta^0)^n)^* E.$$

Thc composition

$$\tau \circ (\delta^0)^n\colon [0] \to [m]$$

is the map that sends 0 to $\tau(n)$. If $\tau(n) = m$, then $\tau \circ (\delta^0)^n = (\delta^0)^m$. Thus we can write $E(\tau) = \mathrm{Id}$. If $\tau(n) < m$ then we have the equalities

$$\tau \circ (\delta^0)^n = (\delta^0)^{\tau(n)} \circ (\delta^1)^{m-\tau(n)},$$

and

$$(\delta^0)^m = (\delta^0)^{\tau(n)} \circ (\delta^1)^{m-\tau(n)-1} \circ \delta^0.$$

In this case we write

$$E(\tau) = B\left((\delta^0)^{\tau(n)} \circ (\delta^1)^{m-\tau(n)-1}\right)^* \alpha.$$

It is easy to see that $E.$ is a simplicial principal G-bundle, that both constructions are functorial and that they determine an equivalence of categories. □

REMARK 2.21. We can make the same definition of principal bundle in the case of topological groups and algebraic groups. Moreover we can give an analogous definition of vector bundle over a simplicial manifold. In the case of simplicial vector bundles, the analogue of Proposition 2.20 also holds.

2.6. The de Rham Algebra of a Simplicial Manifold

The main reference for this section is [**24**, §6]. For a differentiable manifold M, let us denote by $E^*(M, \mathbb{R})$ the de Rham algebra of global differential forms. It is a graded commutative and associative differential algebra. We want to have an analogous object for simplicial differentiable manifolds. Let $M.$ be a simplicial differentiable manifold. Then $E^*(M., \mathbb{R})$ is a simplicial graded commutative associative differential algebra. To this simplicial algebra we can associate a double complex, $\mathcal{C}E^*(M., \mathbb{R})$, and a simple complex denoted by $sCE^*(M., \mathbb{R})$. In this complex we can introduce a multiplicative structure which is associative but only commutative up to homotopy (see Section 7). To remedy this situation, we can construct a differential graded commutative associative algebra that will be called the *simplicial de Rham algebra* (see [**23, 24**]. See also [**54**] for an algebraic analogue).

Let us denote by H^n the hyperplane

$$H^n = \{(x_0, \dots, x_n) \in \mathbb{R}^{n+1} \mid x_0 + \dots + x_n = 1\}.$$

Then $H^{\cdot}$ is a cosimplicial differentiable manifold with faces and degeneracies given by equation (2.3).

DEFINITION 2.22. Let $M.$ be a simplicial differentiable manifold. A *simplicial n-form* over M is a sequence $\varphi = \{\varphi^p\}_p$, where φ^p is a n-form on $H^p \times M_p$, such that, for all $p \geq 0$ and $i = 0, \dots, p$,

$$(\delta^i \times \mathrm{Id})^* \varphi^p = (\mathrm{Id} \times \delta_i)^* \varphi^{p-1},$$

on $H^{p-1} \times M_p$. We will denote by $E^n_{\mathrm{simp}}(M., \mathbb{R})$ the space of all simplicial n-forms. The exterior derivative and the exterior product of forms on $H^p \times X_p$ induce a differential and a commutative and associative product on $E^*_{\mathrm{simp}}(M., \mathbb{R}) = \bigoplus_n E^n_{\mathrm{simp}}(M., \mathbb{R})$ (see [**24**]). We will call $E^*_{\mathrm{simp}}(M., \mathbb{R})$ the *simplicial de Rham algebra* of $M.$.

The complex $E^*_{\mathrm{simp}}(M., \mathbb{R})$ is a bigraded complex, where a p-form φ is said to be of type k, l with $k + l = p$ if $\varphi|_{H^p \times X_p}$ can be written locally as

$$\varphi = \sum a_{I,J} dt_{i_1} \wedge \dots \wedge dt_{i_k} \wedge dx_{j_1} \wedge \dots \wedge dx_{j_l},$$

where $t_0, \dots, t_p$ are baricentric coordinates of H^p and $x_1, \dots x_n$ are local coordinates of X_p.

The complexes $E^*_{\mathrm{simp}}(M., \mathbb{R})$ and $sCE^*(M., \mathbb{R})$ are homotopically equivalent (see [**24**]). In particular the morphism $E^{k,l}_{\mathrm{simp}}(M., \mathbb{R}) \to E^l(M_k, \mathbb{R})$ is obtained by restricting a (k, l)-form to $H^k \times X_k$ and then integrating along the standard simplex $\Delta^k \subset H^k$.

CHAPTER 3

H-Spaces and Hopf Algebras

In the next chapter we will be interested in the homology and cohomology of the general linear group. The product structure of a topological group, or more generally of a H-space, induces a product in homology and a coproduct in cohomology, turning both into Hopf algebras. In this chapter we will review the definition of H-spaces and Hopf algebras and their basic properties. All the results stated are classical and can be found, for instance, in [**46**] or in [**16**].

3.1. Definitions

DEFINITION 3.1. Let (X, e) be a pointed topological space. We say that X is an *H-space* if there is a continuous map $\mu\colon X \times X \to X$, such that, for all $x \in X$, $\mu(x, e) = \mu(e, x) = x$. We say that X is an *associative H-space* if the maps $\mu \circ (\mathrm{Id} \times \mu)$ and $\mu \circ (\mu \times \mathrm{Id})$ from $X \times X \times X$ to X are homotopically equivalent.

Clearly any topological group is an H-space.

Let us fix a commutative ring k. By a graded module we will mean graded by non-negative integers. For any pair of graded k-modules A and B, let $T\colon A \otimes B \to B \otimes A$, be the morphism defined by

$$T(x \otimes y) = (-1)^{\deg x \deg y} y \otimes x.$$

DEFINITION 3.2. A *graded k-algebra* is a graded k-module A together with a unit element $\epsilon\colon k \to A$ and a product $\mu\colon A \otimes A \to A$ such that the compositions

$$A \xrightarrow{\cong} A \otimes k \xrightarrow{\mathrm{Id} \otimes \epsilon} A \otimes A \xrightarrow{\mu} A$$
$$A \xrightarrow{\cong} k \otimes A \xrightarrow{\epsilon \otimes \mathrm{Id}} A \otimes A \xrightarrow{\mu} A$$

are the identity. A graded k-algebra is *associative* if the diagram

$$\begin{array}{ccc} A \otimes A \otimes A & \xrightarrow{\mathrm{Id} \otimes \mu} & A \otimes A \\ {\scriptstyle \mu \otimes \mathrm{Id}} \downarrow & & \downarrow {\scriptstyle \mu} \\ A \otimes A & \xrightarrow{\mu} & A \end{array}$$

is commutative. A graded k-algebra is *commutative* if the diagram

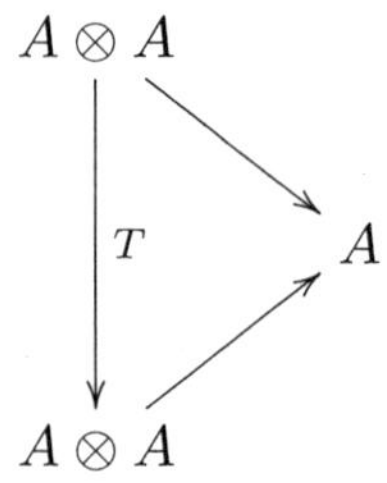

is commutative.

A coalgebra is the dual notion of an algebra.

DEFINITION 3.3. A *graded k-coalgebra* is a graded k-module, A, together with a counit $\eta\colon A \to k$ and a coproduct $\Delta\colon A \to A \otimes A$ such that the compositions

$$
\begin{aligned}
&A \xrightarrow{\Delta} A \otimes A \xrightarrow{\mathrm{Id}\otimes\eta} A \otimes k \xrightarrow{\cong} A \\
&A \xrightarrow{\Delta} A \otimes A \xrightarrow{\eta\otimes\mathrm{Id}} k \otimes A \xrightarrow{\cong} A
\end{aligned}
\tag{3.1}
$$

are the identity. A graded k-coalgebra is *associative* if the diagram

$$
\begin{CD}
A @>\Delta>> A \otimes A \\
@V{\Delta}VV @VV{\mathrm{Id}\otimes\Delta}V \\
A \otimes A @>{\Delta\otimes\mathrm{Id}}>> A \otimes A \otimes A
\end{CD}
$$

is commutative. A graded k-coalgebra is *commutative* if the diagram

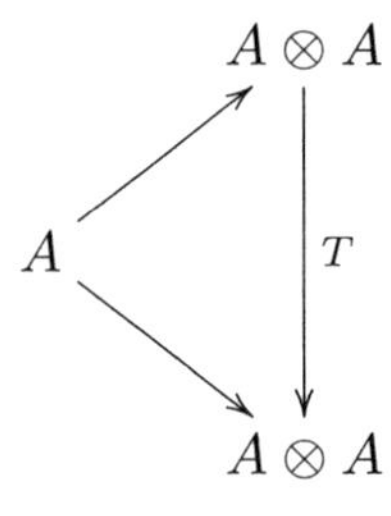

is commutative.

The coproduct is usually called the diagonal map.

Observe that k has a natural structure of k-coalgebra given by the isomorphism $k \to k \otimes k$. Moreover, if A and B are k-coalgebras, there is a natural k-coalgebra structure in $A \otimes B$ with coproduct given by the composition

$$A \otimes B \xrightarrow{\Delta\otimes\Delta} A \otimes A \otimes B \otimes B \xrightarrow{\mathrm{Id}\otimes T\otimes\mathrm{Id}} A \otimes B \otimes A \otimes B.$$

DEFINITION 3.4. A graded k-module A, together with a unit ϵ, a counit η, a product μ and a coproduct Δ, is a *Hopf algebra* if

(1) (A, μ, ϵ) is an associative algebra,
(2) (A, Δ, η) is an associative coalgebra,

(3) μ and ϵ are morphisms of coalgebras and
(4) Δ and η are morphisms of algebras.

Observe that once ϵ and η are morphisms of coalgebras and algebras respectively, the fact that Δ is a morphism of algebras and the fact that μ is a morphism of coalgebras are both equivalent to the commutativity of the following diagram:

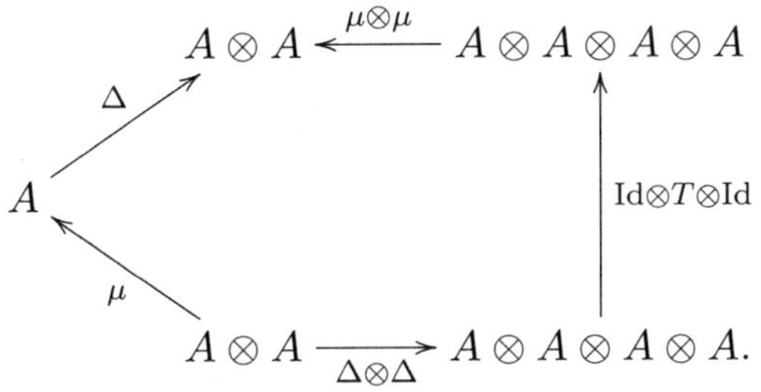

A Hopf algebra A is called *connected* if $\epsilon\colon k \to A_0$ is an isomorphisms. Equivalently A is connected if $\eta\colon A_0 \to k$ is an isomorphism.

EXAMPLE 3.5. Let (X, e) be an associative H-space and let k be a field. Then the diagonal $\Delta\colon X \to X \times X$ and the product structure μ induce a coproduct Δ_* and a product μ_* in the singular homology $H_*(X, k)$. The inclusion $e \to X$ and the projection $X \to e$ induce a unit and a counit respectively. With this structure, $H_*(X, k)$ is a Hopf algebra. Moreover, the coproduct is always commutative. By duality, the singular cohomology is the dual Hopf algebra. Analogously, if we denote by $\overline{H}_*(X)$ the quotient of $H_*(X, \mathbb{Z})$ by its torsion subgroup, then $\overline{H}_*(X)$ is a $\mathbb{Z}$-Hopf algebra with commutative coproduct, and $\overline{H}^*(X)$ is the dual Hopf algebra.

Let us write $I(A) = \operatorname{Ker} \eta$. If A is connected then $I(A) = \bigoplus_{i>0} A_i$. Observe that, since $\eta \circ \epsilon = \mathrm{Id}_K$, we have $A = k \oplus I(A)$ and $I(A) \cong \operatorname{Coker} \epsilon$. Seeing $I(A)$ as a quotient, the coproduct Δ induces a morphism

$$\delta\colon I(A) \to I(A) \otimes I(A).$$

DEFINITION 3.6. Let A be a Hopf algebra. Then *the space of indecomposable elements*, denoted $Q(A)$, is the cokernel of the morphisms

$$\mu\colon I(A) \otimes I(A) \to I(A).$$

The *space of primitive elements*, denoted $P(A)$, is the kernel of the morphism

$$\delta\colon I(A) \to I(A) \otimes I(A).$$

Observe that, since (A, Δ, η) is a coalgebra, by (3.1), for any element $a \in I(A)$, we have

$$\Delta(a) = 1 \otimes a + \delta(a) + a \otimes 1.$$

Thus, $a \in I(A)$ is primitive if and only if $\Delta(a) = 1 \otimes a + a \otimes 1$.

3.2. Some Examples

Let us give some examples of coalgebras and Hopf algebras over $\mathbb{Z}$.

EXAMPLE 3.7. Let $\bigwedge(x_1, \dots, x_n)$ be the exterior algebra generated by the elements $x_1, \dots, x_n$ of odd degree. We can define a Hopf algebra structure imposing that the elements x_i are primitive elements. The dual of this Hopf algebra is again the exterior algebra generated by the primitive elements $y_1, \dots, y_n$, where (y_i) is the dual basis of (x_i).

By the Samelson–Leray theorem (see Theorem 3.15) any torsion free commutative $\mathbb{Z}$-Hopf algebra generated by elements of odd degree is isomorphic (as Hopf algebra) to an exterior algebra generated by primitive elements.

Let us give a pair of examples of evenly generated Hopf algebras.

EXAMPLE 3.8. Let $A = \mathbb{Z}[x]$ be the polynomial ring in one variable of degree 2. Then its dual coalgebra is, as graded abelian group

$$A^* = \bigoplus_{i \geq 0} \mathbb{Z}\gamma_i,$$

where γ_i has degree $2i$ and is the dual of x^i. The coproduct is given by

$$\Delta\gamma_i = \sum_{j+k=i} \gamma_j \otimes \gamma_k.$$

For instance, we can define a Hopf algebra structure in A by imposing that x is primitive; then the algebra structure of A^* satisfies

$$\gamma_i = \frac{\gamma_1^i}{i!}.$$

Thus it is isomorphic to the divided power polynomial algebra $\Gamma[\gamma_1]$, with γ_1 primitive. Observe that A and A^* are not isomorphic. The former is generated by its primitive part and the latter is not.

The following example of self-dual Hopf algebra (see [**51**]) is more interesting.

EXAMPLE 3.9. Let B be the Hopf algebra such that, as an algebra it is the polynomial ring $\mathbb{Z}[b_1, b_2, \dots]$, with b_i of degree $2i$. And with a coproduct given by

$$\Delta b_i = \sum_{j+k=i} b_j \otimes b_k. \tag{3.2}$$

The dual Hopf algebra, $B^\vee$ has the algebra structure of the polynomial ring $\mathbb{Z}[y_1, y_2, \dots]$, where y_i is the dual of b_1^i. Moreover, the coproduct structure is also given by

$$\Delta y_i = \sum_{j+k=i} y_j \otimes y_k.$$

Thus this Hopf algebra is self dual.

In this example we can obtain an inductive formula for the primitive elements of the coalgebra algebra $B^\vee$ (see [**51**]). Let us write

$$\begin{aligned} \mathrm{pr}_1 &= y_1, \\ \mathrm{pr}_n &= (-1)^{n+1} n y_n + \sum_{j=1}^{n-1} (-1)^{j+1} y_j \mathrm{pr}_{n-j}, \quad \text{for } n > 1. \end{aligned} \tag{3.3}$$

PROPOSITION 3.10. *The elements* pr_i *form a basis of* $P(B)$*. Moreover,*

$$\langle \mathrm{pr}_i, b_i \rangle = 1.$$

3.3. The Structure of Hopf Algebras

The presence of two compatible operations imposes many restrictions on the structure of Hopf algebras. We will recall some classical results in this direction. For simplicity, we will state most of the results for Hopf algebras over a field of characteristic zero or for torsion free Hopf algebras over the ring of integers.

The first result in the study of the structure of Hopf algebras is the following.

PROPOSITION 3.11. *Let A be a connected Hopf algebra over a field of characteristic zero.*

(1) *The product is commutative if and only if the natural morphism*

$$P(A) \to Q(A)$$

is a monomorphism.

(2) *The coproduct is commutative if and only if the natural morphism*

$$P(A) \to Q(A)$$

is an epimorphism. In particular, a connected Hopf algebra with commutative coproduct is generated, as an algebra, by the space of primitive elements.

The first statement remains valid if A is a torsion free $\mathbb{Z}$-Hopf algebra. But as Example 3.8 shows, the second statement does not remain true in this case.

From now on, we fix a field k of characteristic zero and a connected Hopf k-algebra A with commutative coproduct.

We can define a structure of graded Lie algebra on A writing

$$[x, y] = xy - (-1)^{\deg x \deg y} yx.$$

It is easy to see that the space of primitive elements, $P(A)$, is a Lie subalgebra of A. Let us denote by $U(P(A))$ the universal enveloping algebra of $P(A)$. Since the inclusion $P(A) \to A$ is a morphism of the Lie algebra $P(A)$ into an associative algebra A, there is a unique extension to a morphism $U(P(A)) \to A$. Moreover, there is a natural structure of Hopf algebra on $U(P(A))$ (see [**46**] for details). Then the main structure theorem is

THEOREM 3.12. *Let A be a connected Hopf algebra with commutative coproduct over a field of characteristic zero. Then the natural map $U\big(P(A)\big) \to A$ is an isomorphism of Hopf algebras.*

From this theorem and the Poincaré–Birkhoff–Witt Theorem we can completely determine the structure of k-module of A.

For a graded k-module B, let us denote by $\bigwedge(B)$ the free graded commutative and associative algebra generated by B. This is the exterior algebra over the odd subspace of B tensored with the symmetric algebra over the even subspace. Let L be a graded Lie algebra. Let us denote by $L^\sharp$ the Lie algebra with the same underlying module as L but with abelian Lie product. Then $\Lambda(L) = U(L^\sharp)$.

Using the Lie bracket, we can define a filtration F on $U(L)$ and on $\bigwedge(L) = U(L^\sharp)$. This filtration is called the Lie filtration.

THEOREM 3.13 (Poincaré–Birkhoff–Witt). *Let L be a graded Lie k-algebra. Then there is a natural isomorphism of bigraded Hopf algebras*

$$\mathrm{Gr}_F\big(U(L^\sharp)\big) \to \mathrm{Gr}_F\big(U(L)\big).$$

COROLLARY 3.14. *Let L be a graded Lie k-algebra. Then there is a (nonnatural) isomorphism of k-modules between $\bigwedge(L)$ and $U(L)$.*

In the case when a Hopf algebra is generated by its odd part we have a more precise statement, the Samelson-Leray theorem.

THEOREM 3.15 (Samelson–Leray). *Let A be a torsion free, connected Hopf algebra over $\mathbb{Z}$, such that the product is commutative and $Q_n(A)$ is torsion for n even. Then*

(1) *$Q(A)$ is torsion free.*
(2) *The morphism $P(A) \to Q(A)$ is an isomorphism.*
(3) *The coproduct is commutative.*
(4) *The natural morphism $\bigwedge\big(P(A)\big) \to A$ is an isomorphism of Hopf algebras.*

We can apply the structure theorems to the homology and cohomology of compact H-spaces. The compacity implies that the homology algebra $H_*(X, \mathbb{Q})$ is finite dimensional. Thus by Theorem 3.13 we obtain that $P_n\big(H_*(X, \mathbb{Q})\big) = 0$ for n even. In consequence $Q^n\big(\overline{H}^*(X)\big)$ is torsion for n even. Thus the Samelson–Leray theorem implies

PROPOSITION 3.16. *Let X be a compact H-space. Then $\overline{H}^*(X)$ and $\overline{H}_*(X)$ are, as Hopf algebras, isomorphic to an exterior algebra generated by primitive elements of odd degree.*

3.4. Rational Homotopy of H-Spaces

As we noted above, the main example of connected Hopf algebra with commutative coproduct is the singular homology of a connected H-space. In this case we want to give a more geometric interpretation of the space of primitive elements. Let (X, e) be a connected associative H-space. Let us write $P_n(X, \mathbb{Q}) = P_n\big(H_*(X, \mathbb{Q})\big)$.

Let S^n be the n-dimensional sphere. Observe that the elements of the cohomology of the sphere are necessarily primitive elements. This implies that the Hurewicz morphism factorizes as

$$\pi_n(X, e) \xrightarrow{\lambda} P_n(X, \mathbb{Q}) \to H_n(X, \mathbb{Q}).$$

Moreover, one can define a Lie product in $\pi_*(X, e)$, called the *Samelson product*, such that λ is a morphism of Lie algebras. For the proof of the following result see for instance [**49**].

THEOREM 3.17 (Cartan–Serre). *Let (X, e) be a path-wise connected associative H-space. Then the Hurewicz map induces an isomorphism of Lie algebras*

$$\lambda\colon \pi_*(X, e) \otimes \mathbb{Q} \to P_*(X, \mathbb{Q}).$$

Therefore we obtain an isomorphism of Hopf algebras

$$U(\lambda)\colon U(\pi_*(X, e) \otimes \mathbb{Q}) \to H_*(X, \mathbb{Q}).$$

CHAPTER 4

The Cohomology of the General Linear Group

4.1. The General Linear Group and the Stiefel Manifolds

In this section we shall compute the singular cohomology of the complex general linear group, $\mathrm{GL}_n(\mathbb{C})$ with integral coefficients. Since $\mathrm{GL}_n(\mathbb{C})$ is homotopically equivalent to the unitary group U_n it is enough to compute the cohomology ring of this latter group. From the last section we know that $\overline{H}^*(\mathrm{U}_n)$ is an exterior algebra generated by elements of odd degree. Our objective now is to show that $H^*(\mathrm{U}_n, \mathbb{Z})$ is torsion free and obtain a set of canonical generators. These cohomology groups will be computed by induction using Stiefel manifolds. The computations are classical and can be found, for instance, in [**60**].

We will consider the set of groups $\{\mathrm{U}_n\}_n$ as a directed system with morphisms $\varphi_{n,m} \colon \mathrm{U}_m \to \mathrm{U}_n$, for $m \leq n$, given by

$$\varphi_{n,m}(A) = \begin{pmatrix} A & 0 \\ 0 & I \end{pmatrix}.$$

Usually, we will identify U_m with its image in any of the groups U_n, for $n \geq m$.

DEFINITION 4.1. For any pair of integers $0 \leq l \leq n$, the *Stiefel manifold*, $V_{n,l}$, is defined as

$$V_{n,l} = \mathrm{U}_n \,/\, \mathrm{U}_{n-l}\,.$$

Geometrically $V_{n,l}$ can be interpreted as the set of sequences of l orthonormal vectors in $\mathbb{C}^n$.

Observe that

$$V_{n,n} = \mathrm{U}_n, \quad \text{and} \quad V_{n,1} = S^{2n-1},$$

the $2n-1$ dimensional sphere. Moreover the natural map $V_{n,l+1} \to V_{n,l}$ is a fibre bundle with fibres $\mathrm{U}_{n-l} \,/\, \mathrm{U}_{n-l-1} \cong S^{2n-1-2l}$.

THEOREM 4.2. *The ring $H^*(V_{n,l}, \mathbb{Z})$ is an exterior algebra generated by elements x_j in degree $2j-1$, for $n-l < j \leq n$.*

PROOF. The proof of the theorem is done by induction over l. For $l = 1$ it is true because we have $V_{n,1} = S^{2n-1}$. By the induction hypothesis we may assume that it is true for $V_{n,l}$. Let us consider the fibre bundle $V_{n,l+1} \to V_{n,l}$ with fibre $F = S^{2n-1-2l}$.

LEMMA 4.3. *The Leray spectral sequence of the fibre bundle*

$$V_{n,l+1} \to V_{n,l}$$

has E_2 *term*

$$E_2^{p,q} = H^p(V_{n,l}, \mathbb{Z}) \otimes H^q(S^{2n-1-2l}, \mathbb{Z}).$$

PROOF. Let us write F, E and B for the fibre, the total space and the base of the fibre bundle respectively. Let us choose a point $b \in B$. The E_2 term of the Leray spectral sequence is given by

$$E_2^{p,q} = H^p\big(B, \mathcal{H}^q(F, \mathbb{Z})\big).$$

The fibre bundle $V_{n,l+1} \to V_{n,l}$ is the quotient of the principal U_{n-l}-bundle $\mathrm{U}_n \to V_{n,l}$ by the closed subgroup U_{n-l-1}. Therefore, the group U_{n-l} acts continuously on the fibre, and the transition functions have values in this group. Hence, the monodromy of an element $\gamma \in \pi_1(B, b)$, is given by an element of $f_\gamma \in \mathrm{U}_{n-l}$. But since this group is connected, this action is homotopically equivalent to the identity. Thus the local system $\mathcal{H}^*(F, \mathbb{Z})$ is trivial. Moreover, since by the induction hypothesis the cohomology of the base is a finitely generated free abelian group, we obtain the result. □

As a consequence of the above lemma, $E_2^{p,q}$ is zero for $q \neq 0, 2n-1-2l$. Hence the only differential that may be different from zero is d_{2n-2l}. Since E_{2n-2l} is an algebra and d_{2n-2l} is a derivation, this differential is determined by

$$d_{2n-2l} \colon E_{2n-2l}^{0,2n-2l-1} = H^{2n-2l-1}(F, \mathbb{Z}) \to E_{2n-2l}^{2n-2l,0} = H^{2n-2l}(B, \mathbb{Z}).$$

But by the induction hypothesis this last group is zero. Therefore $E_\infty = E_2$. Using again the induction hypothesis and Lemma 4.3 we obtain that E_∞ is an exterior algebra generated by elements v_j in degree $2j-1$, for $n-l-1 < j \leq n$. Therefore Theorem 4.2 is a consequence of the following result.

LEMMA 4.4. *Let* A *be a finitely generated graded commutative algebra over* $\mathbb{Z}$ *and let* F *be a homogeneous decreasing filtration such that* $F^i \cdot F^j \subset F^{i+j}$. *Let* GA *be the associated bigraded algebra:*

$$GA^{k,l} = F^k A^{k+l} / F^{k+1} A^{k+l}.$$

If GA *is an exterior algebra generated by* r *bi-homogeneous elements of odd total degree, then* A *is an exterior algebra generated by* r *elements of the same total degree.*

PROOF. Let $\{v_1, \dots, v_r\}$ be a set of generators of GA with $v_j \in GA^{k_j, l_j}$. For each v_j let us choose a representative $x_j \in F^{k_j} A^{k_j + l_j}$. Let B be the exterior algebra generated by symbols u_j. By the universality of the exterior algebra there is a natural morphism $\varphi \colon B \to A$ that sends u_j to x_j, for $j = 1, \dots, r$. Moreover, the filtration F induces a multiplicative filtration on B and φ becomes a filtered morphism. Since the graded morphism $\operatorname{Gr} \varphi$ is an isomorphism, then φ is also an isomorphism. This completes the proof of the Lemma and of Theorem 4.2. □

Let us choose a set of distinguished generators of $H^*(\mathrm{U}_n, \mathbb{Z})$. To this end we choose a square root of -1. Thus an orientation of $\mathbb{C}^n$. Let $\sigma_{2n-1} \in H^{2n-1}(S^{2n-1}, \mathbb{Z})$ be the class determined by this orientation. Let us denote by $\pi_n \colon \mathrm{U}_n \to \mathrm{U}_n / \mathrm{U}_{n-1} = S^{2n-1}$ the morphism that sends a unitary matrix to its last column.

DEFINITION 4.5. Let $\alpha_{n,2p-1} \in H^{2p-1}(\mathrm{U}_n, \mathbb{Z})$ be the elements determined inductively as follows

(1) $\alpha_{n,2n-1} = \pi_n^*(\sigma_{2n-1})$.
(2) $\varphi_{n,n-1}^*(\alpha_{n,2p-1}) = \alpha_{n-1,2p-1}$, for $1 \leq p < n$.

Observe that in this definition we are using the fact that

$$\varphi_{n,n-1}^* \colon H^j(\mathrm{U}_n, \mathbb{Z}) \to H^j(\mathrm{U}_{n-1}, \mathbb{Z})$$

is an isomorphism for $j < 2n - 1$. Usually we will denote the generators $\alpha_{n,2p-1}$ for $n \geq p$ simply by α_{2p-1}.

REMARK 4.6. The elements α_{2p-1} are primitive for the Hopf algebra structure of $H^*(\mathrm{U}_n, \mathbb{Z})$ (see Section 4.2). Since $P^{2p-1}(\mathrm{U}_n, \mathbb{Z})$ is a free abelian group of rank onc, this determines α_{2p-1} up to the sign.

REMARK 4.7. By duality, the homology algebra of U_n is also an exterior algebra generated by one element in each degree $2p-1$ for $p = 1, \dots, n$. Moreover the generators α_{2p-1} determine a set of generators $\beta_{2p-1} \in H_{2p-1}(\mathrm{U}_n, \mathbb{Z})$ which are also primitive elements.

4.2. Classifying Spaces and Characteristic Classes

Let G be a Lie group. A universal principal G-bundle is a principal G-bundle (r, E, π, B), such that the total space E, is contractible. The base is called a classifying space for the group G (see, for instance, [**52**]). Any two classifying spaces for a given group G are homotopically equivalent. The universality is given by the following property: For any topological principal G-bundle (r, F, π, X), with X paracompact, there is a continuous function $f \colon X \to B$ such that $F = f^*E$. Moreover, the function f is determined up to homotopy.

It is easy to construct B as a topological space. But if we want more structure (differentiable, algebraic, ...) it is more interesting to use simplicial objects. Recall that in Example 2.5 we defined a contractible simplicial topological space $E.G$. We can define a right action of G on $E.G$ by

$$(g_0, \dots, g_k)g = (g_0 g, \dots, g_k g).$$

This action is called the diagonal right action and it commutes with the faces and degeneracies. Thus the quotient is a simplicial differentiable manifold.

DEFINITION 4.8. Let G be a Lie group. The *classifying space* $B.G$ of G is the quotient of $E.G$ by the diagonal right action. Thus it is the simplicial differentiable manifold given by

$$BG_k = \overbrace{G \times \cdots \times G}^{k},$$

$$\begin{aligned}
\delta^0(g_1, \dots, g_k) &= (g_2, \dots, g_k), \\
\delta^i(g_1, \dots, g_k) &= (g_1, \dots, g_i g_{i+1}, \dots, g_k), && \text{for } i = 1, \dots, k-1, \\
\delta^k(g_1, \dots, g_k) &= (g_1, \dots, g_{k-1}), \\
\sigma^i(g_1, \dots, g_k) &= (g_1, \dots, g_i, 1, g_{i+1}, \dots, g_k), && \text{for } i = 0, \dots, k.
\end{aligned}$$

Then $E.G$ is an universal principal G-bundle over $B.G$ and the morphism $E.G \to B.G$ is given by

$$(g_0, \dots, g_k) \mapsto (g_0 g_1^{-1}, \dots, g_{k-1} g_k^{-1}).$$

The main objective of this section is to recall the structure of the cohomology of the classifying space $B.\mathrm{GL}_n(\mathbb{C})$. The Chern classes of complex vector bundles and the cohomology of the classifying space $B.\mathrm{GL}_n(\mathbb{C})$, are two topics intimately related. One can, as in [**24**], compute the cohomology of the classifying space and use it to define characteristic classes or, as in [**47**], one can introduce first characteristic classes and use them to study the cohomology of the classifying space. We will follow the second line of thought. As in the previous section, we can use the compact group U_n instead of the group $\mathrm{GL}_n(\mathbb{C})$.

Let us recall a definition of Chern classes. Let X be a differentiable manifold (simplicial or not) and let $\pi\colon F \to X$ be a complex vector bundle of rank n. Let $\pi_S\colon S \to X$ be the associated S^{2n-1}-bundle. Let $E_r^{p,q}$ be the Leray spectral sequence of the bundle S. Since it is a S^{2n-1}-bundle, then $E_r^{p,q} = 0$ for $q \neq 0, 2n-1$. Thus the only nonzero differential is d_{2n}. As in the previous section, the standard orientation of F as complex vector bundle defines a class $\sigma_{2n+1} \in H^{2n-1}(S^{2n-1}, \mathbb{Z}) = E_2^{0,2n-1}$. The Euler class of F is the class $e(F) = d_{2n}\sigma_{2n-1} \in E^{2n,0} = H^{2n}(X, \mathbb{Z})$.

To give an inductive definition of Chern classes we need two more facts. First, observe that, for $j < 2n-1$, the morphism

$$\pi_S^*\colon H^j(X, \mathbb{Z}) \to H^j(S, \mathbb{Z})$$

is an isomorphism. The other fact is that the vector bundle $\pi_S^* F$ has a canonical rank one trivial subbundle L. The fibre of L over a point v is the line spanned by v. Let us write $F_0 = \pi_S^* F / L$.

DEFINITION 4.9. Let F be a rank n vector bundle over X. The *integer valued Chern classes* of F, $b_p(F) \in H^{2p}(X, \mathbb{Z})$ are determined inductively by the following conditions:

(1) $b_n(F) = e(F)$.
(2) For $p < n$,

$$\pi_S^* b_p(F) = b_p(F_0).$$

(3) For $p > n$, $b_p(F) = 0$.

For our purposes, it will be convenient to define also the twisted Chern classes, which differ from the integer valued Chern classes by a normalization factor. For any subgroup Λ of $\mathbb{C}$ we will write

$$\Lambda(p) = (2\pi i)^p \Lambda \subset \mathbb{C}.$$

DEFINITION 4.10. The *twisted Chern classes* are

$$c_p(F) = (2\pi i)^p b_p(F) \in H^{2p}\big(X, \mathbb{Z}(p)\big).$$

From the universal principal G-bundle $E.\mathrm{U}_n \to B.\mathrm{U}_n$ we can define a universal vector bundle. Let us denote by $\sim$ the equivalence relation on $E.\mathrm{U}_n \times \mathbb{C}^n$ given by

$$(xg, v) \sim (x, gv), \quad \text{for all } x \in E.\mathrm{U}_n,\ v \in \mathbb{C}^n \text{ and } g \in \mathrm{U}_n.$$

DEFINITION 4.11. The *universal rank n vector bundle*, is the vector bundle $F_n \to B.\mathrm{U}_n$ defined by

$$F_n = E.\mathrm{U}_n \underset{\mathrm{U}_n}{\times} \mathbb{C}^n = E.\mathrm{U}_n \times \mathbb{C}^n / \sim$$

For the proof of the following theorem see [**47**].

THEOREM 4.12. *The ring $H^*(B\,\mathrm{U}_n, \mathbb{Z})$ is a polynomial ring generated by the elements $b_p(F_n)$, $p = 1, \dots, n$, with $b_p(F_n)$ of degree $2p$.*

4.3. The Suspension

In Section 4.1 we described a set of canonical generators of $H^*(\mathrm{U}_n, \mathbb{Z})$ and in Section 4.2 we recalled that a set of canonical generators of $H^*(B.\mathrm{U}_n, \mathbb{Z})$ are given by the Chern classes of the universal bundle. The aim of this section is to show the relationship between the two sets of generators. This relationship is given by the suspension map. The main references for this section are [**5, 53**].

Let (E, π, B) be a fibre bundle with B connected. Let us choose a point $x \in B$, let F be the fibre at the point x, and let $i \colon F \to E$ be the inclusion. Let $j > 0$ be an integer and let $[\alpha] \in H^j(B, \mathbb{Z})$ be a class such that $\pi^*[\alpha] = 0$. Let $\alpha \in C^j(B, \mathbb{Z})$ be a representative of $[\alpha]$. By hypothesis, the cochain $\pi^*(\alpha)$ is exact. Let us choose any cochain $\beta \in C^{j-1}(E, \mathbb{Z})$ such that $d\beta = \pi^*(\alpha)$. Since the morphism $\pi \circ i \colon F \to B$ factorizes through the point x, the cochain $i^*\pi^*(\alpha) = 0$. Thus $i^*(\beta)$ is closed. Moreover it is easy to see that the cohomology class $[i^*(\beta)]$ only depends on the class $[\alpha]$.

DEFINITION 4.13. The *suspension* of $[\alpha]$ is the class $\mathfrak{s}[\alpha] = [i^*(\beta)]$.

Thus the suspension is a morphism from $\operatorname{Ker} \pi^*$ to $H^*(F, \mathbb{Z})$.

Let us denote by $T^*(F, \mathbb{Z})$ the image of $\mathfrak{s}$. The elements of this group are called *transgressive*. Let $L^*(B, \mathbb{Z})$ denote the kernel of $\mathfrak{s}$. Thus the suspension gives us an isomorphism

$$\mathfrak{s} \colon \frac{\operatorname{Ker} \pi^*}{L^*(B, \mathbb{Z})} \to T^*(F, \mathbb{Z}).$$

The inverse of this isomorphism is called the *transgression*. Observe that, composing with the inclusion $\operatorname{Ker}\pi^* \subset H^*(B,\mathbb{Z})$ we may assume that the transgression is a morphism

$$\mathsf{t}\colon T^{j-1}(F,\mathbb{Z}) \to \frac{H^j(B,\mathbb{Z})}{L^j(B,\mathbb{Z})}.$$

We recall another description of the transgression (for a proof and more details see [**5**, §5] and [**45**, §6.1]). Let $E_r^{p,q}$ be the Leray spectral sequence for the fibre bundle $\pi\colon E \to B$. Let us assume that the local systems $\mathcal{H}^j(F,\mathbb{Z})$, $j \geq 0$, are trivial. This hypothesis will be satisfied in all the examples. Then we may identify $H^j(F,\mathbb{Z})$ with $E_2^{0,j}$ by the morphism i^*. Let us identify also $H^{j+1}(B,\mathbb{Z})$ with $E_2^{j+1,0}$ by the morphism π_*. Let us denote by κ_{j+1}^2 the projection $E_2^{j+1,0} \to E_{j+1}^{j+1,0}$.

PROPOSITION 4.14. *There is a commutative diagram*

$$\begin{array}{ccccc}
T^j(F,\mathbb{Z}) & \xrightarrow{\ \mathsf{t}\ } & \frac{H^{j+1}(B,\mathbb{Z})}{L^{j+1}(B,\mathbb{Z})} & \longleftarrow & H^{j+1}(B,\mathbb{Z}) \\
{\scriptstyle i^*}\big\uparrow & & {\scriptstyle \pi^*}\big\downarrow & & {\scriptstyle \pi^*}\big\downarrow \\
E_{j+1}^{0,j} & \xrightarrow{\ d_{j+1}\ } & E_{j+1}^{j+1,0} & \xleftarrow{\ \kappa_{j+1}^2\ } & E_2^{j+1,0},
\end{array}$$

where the vertical arrows are isomorphisms. In particular, the transgressive elements, $T^j(F,\mathbb{Z})$, are those in the successive kernels of the morphisms $d_2,\dots,d_j$ and $L^{j+1}(B,\mathbb{Z})$ is the kernel of κ_{j+1}^2.

REMARK 4.15. If $E_r^{p,q}$ is a first quadrant spectral sequence, then $E_{j+1}^{0,j}$ is a subobject of $E_2^{0,j}$, and $E_{j+1}^{j+1,0}$ is a quotient of $E_2^{j+1,0}$. The morphism

$$d_{j+1}\colon E_{j+1}^{0,j} \to E_{j+1}^{j+1,0}$$

is called the transgression of this spectral sequence. Thus the meaning of Proposition 4.14 is that the transgression we have defined as the inverse of the suspension agrees with the transgression for the Leray spectral sequence of the fibre bundle.

EXAMPLE 4.16. Let G be a connected Lie group and let (r, E, π, B) be a principal G-bundle, with fibre F. Then, for all j, the local systems $\mathcal{H}^j(F,\mathbb{Z})$ are trivial, and we may identify $H^j(F,\mathbb{Z})$ with $H^j(G,\mathbb{Z})$. If E is a universal principal G-bundle, then it is contractible. Therefore the suspension gives us a morphism

$$H^j(B,\mathbb{Z}) \to H^{j-1}(G,\mathbb{Z}).$$

Proposition 4.14 has the following consequence.

PROPOSITION 4.17. *Let $\alpha_1, \alpha_3, \ldots, \alpha_{2n-1}$ be the generators of the group $H^*(\mathrm{U}_n, \mathbb{Z})$ introduced in Definition 4.5. Then the elements α_{2p-1} are transgressive in the universal principal U_n-bundle. Moreover, if $\mathfrak{s}$ is the suspension, F_n is the universal rank n vector bundle, and $b_p(F_n)$ are the integer valued Chern classes, then $\mathfrak{s}\big(b_p(F_n)\big) = \alpha_{2p-1}$.*

PROOF. Since the generators α_{2p-1}, the Chern classes of the universal bundle and the transgression are natural for the morphisms $\varphi_{n,m}\colon \mathrm{U}_m \to \mathrm{U}_n$, it is enough to show that α_{2n-1} is transgressive and that $\mathfrak{t}(\alpha_{2n-1})$ is equal to $b_n(F_n)$ modulo elements of $L^{2n}(B.\mathrm{U}_n, \mathbb{Z})$. Let us denote by S_n the sphere bundle over $B.\mathrm{U}_n$ associated to the universal vector bundle F_n. Let $v_0 = (0, \ldots, 0, 1)^t \in \mathbb{C}^n$. Let $\epsilon\colon E.\mathrm{U}_n \to F_n$ be the morphism that sends a point $x \in E.\mathrm{U}_n$ to the class of (x, v_0). Clearly the image of ϵ lies in S_n. Moreover, if we restrict ϵ to $E_0\,\mathrm{U}_n = \mathrm{U}_n$, we obtain the morphism $\pi_n\colon \mathrm{U}_n \to S^{2n-1}$. Let $\sigma_{2n-1} \in H^{2n-1}(S^{2n-1}, \mathbb{Z})$ be the generator determined by the orientation of $\mathbb{C}^n$. By definition σ_{2n-1} is transgressive in the fibration S_n and $\mathfrak{t}(\sigma_{2n-1}) = b_n(F_n)$. Since $\epsilon^*(\sigma_{2n-1}) = \alpha_{2n-1}$ and ϵ^* induces a morphism of spectral sequences we obtain the result. □

The next fact we will recall is the relationship between the suspension and the spaces of primitive and indecomposable elements. Proofs of the following results can be found in [**50, 53**].

THEOREM 4.18. *Let (r, E, π, B) be a principal G-bundle with E acyclic. Then the suspension has the following properties.*

1. $T^*(G, \mathbb{Z}) \subset P^*(G, \mathbb{Z})$
2. *The kernel of the morphism $H^*(B, \mathbb{Z}) \to Q^*(B, \mathbb{Z})$ is contained in $L^*(B, \mathbb{Z})$*

Therefore the suspension induces a morphism

$$\mathfrak{s}\colon Q^*(B, \mathbb{Z}) \to P^{*-1}(G, \mathbb{Z}).$$

In the case of the universal bundle for $\mathrm{GL}_n(\mathbb{C})$ we can apply a more precise result due to A. Borel.

THEOREM 4.19. *Let (r, E, π, B) be a principal G-bundle with E acyclic and such that $H^*(G, \mathbb{Z}) = \bigwedge(V)$, where V is an odd graded module. Then the suspension induces an isomorphism $Q^j(B, \mathbb{Z}) \to P^{j-1}(G, \mathbb{Z})$. Moreover $H^*(B, \mathbb{Z}) = \bigwedge(E[-1])$.*

Observe that, as a corollary of Theorem 4.19 and of Proposition 4.17 we obtain that the generators $\{\alpha_{2p-1}\}_{p=1,\ldots,n}$ of $H^*(\mathrm{GL}_n(\mathbb{C}), \mathbb{Z})$ are primitive.

4.4. The Stability of Homology and Cohomology

The groups $\mathrm{GL}_n(\mathbb{C})$ form a directed system as in Section 4.1. Let us write

$$\mathrm{GL}(\mathbb{C}) = \varinjlim \mathrm{GL}_n(\mathbb{C}).$$

Then $\mathrm{GL}(\mathbb{C})$ is a topological group with the limit topology. Moreover, since any compact subset of $\mathrm{GL}(\mathbb{C})$ is contained in some $\mathrm{GL}_n(\mathbb{C})$ we have that

$$H_j(\mathrm{GL}(\mathbb{C}),\mathbb{Z}) = \varinjlim H_j(\mathrm{GL}_n(\mathbb{C}),\mathbb{Z}),$$
$$H^j(\mathrm{GL}(\mathbb{C}),\mathbb{Z}) = \varprojlim H^j(\mathrm{GL}_n(\mathbb{C}),\mathbb{Z}).$$

If $m < n$, the morphisms

$$(\varphi_{n,m})_* \colon H_j(\mathrm{GL}_m(\mathbb{C}),\mathbb{Z}) \to H_j(\mathrm{GL}_n(\mathbb{C}),\mathbb{Z}),$$
$$(\varphi_{n,m})^* \colon H^j(\mathrm{GL}_n(\mathbb{C}),\mathbb{Z}) \to H^j(\mathrm{GL}_m(\mathbb{C}),\mathbb{Z})$$

are isomorphisms for $j \leq 2m$. Therefore, for $j \leq 2m$, the morphisms

$$(\varphi_m)_* \colon H_j(\mathrm{GL}_m(\mathbb{C}),\mathbb{Z}) \to H_j(\mathrm{GL}(\mathbb{C}),\mathbb{Z}),$$
$$(\varphi_m)^* \colon H^j(\mathrm{GL}(\mathbb{C}),\mathbb{Z}) \to H^j(\mathrm{GL}_m(\mathbb{C}),\mathbb{Z})$$

are isomorphisms. This result is called the *stability of the homology and cohomology of the general linear group*. All the classical series of Lie groups enjoy a similar property.

By the results of Section 4.1 we obtain that, as Hopf algebras,

$$H^*(\mathrm{GL}(\mathbb{C}),\mathbb{Z}) = \bigwedge(\alpha_1,\alpha_3,\dots), \tag{4.1}$$
$$H_*(\mathrm{GL}(\mathbb{C}),\mathbb{Z}) = \bigwedge(\beta_1,\beta_3,\dots), \tag{4.2}$$

where the elements α_{2p-1}, β_{2p-1}, $p = 1,2,\dots$ are primitive of degree $2p-1$.

There is also a similar stability result for the homology and cohomology of the classifying space $B.\mathrm{GL}(\mathbb{C})$. For the cohomology we obtain

$$H^*(B.\mathrm{GL}(\mathbb{C}),\mathbb{Z}) = \mathbb{Z}[b_1,b_2,\dots], \tag{4.3}$$

where the b_i are the Chern classes of the universal bundle. By duality, the homology is

$$H_*(B.\mathrm{GL}(\mathbb{C}),\mathbb{Z}) = \mathbb{Z}[y_1,y_2,\dots], \tag{4.4}$$

where y_i has degree $2i$ and is the dual of b_1^i. The coalgebra structure of the homology is given by

$$\Delta(y_i) = \sum_{j+k=i} y_j \otimes y_k.$$

As in Example 3.9 a basis of the primitive elements in homology is given by

$$\begin{aligned} \mathrm{pr}_1 &= y_1, \\ \mathrm{pr}_n &= (-1)^{n+1} n y_n + \sum_{j=1}^{n-1} (-1)^{j+1} y_j \mathrm{pr}_{n-j}, \quad \text{for } n > 1. \end{aligned}$$

By Proposition 4.17 we have

PROPOSITION 4.20. (1) *The suspension*

$$\mathfrak{s} \colon H^{2p}(B.\mathrm{GL}(\mathbb{C}),\mathbb{Z}) \to H^{2p-1}(\mathrm{GL}(\mathbb{C}),\mathbb{Z})$$

is given by

$$\mathfrak{s}(b_p) = \alpha_{2p-1}.$$

(2) *The suspension map in homology*:

$$\mathfrak{s}^\vee : H_*(\mathrm{GL}(\mathbb{C}), \mathbb{Z}) \to H_*(B.\mathrm{GL}(\mathbb{C}), \mathbb{Z}),$$

is given by

$$\mathfrak{s}^\vee(\beta_{2p-1}) = \mathrm{pr}_p.$$

It is clear that the space $B.\mathrm{GL}_n(\mathbb{C})$ cannot have a structure of H-space, because its cohomology ring is finite dimensional and evenly generated. On the other hand, the infinite classifying space $B.\mathrm{GL}(\mathbb{C})$ has a structure of H-space (see [**22**]). This implies that the homology and cohomology of $B.\mathrm{GL}(\mathbb{C})$ have a structure of Hopf algebras. Moreover these Hopf algebras are isomorphic to the Hopf algebra of Example 3.9. That is, the coproduct in cohomology is given by

$$\Delta b_i = \sum_{j+k=i} b_j \otimes b_k,$$

and the product in homology is determined by the equality (4.4) being an algebra isomorphism. Therefore, as in Example 3.9, a basis of the primitive elements of the cohomology is determined inductively by

$$\begin{aligned} \mathrm{pr}^1 &= b_1, \\ \mathrm{pr}^n &= (-1)^{n+1} n b_n + \sum_{j=1}^{n-1} (-1)^{j+1} b_j \mathrm{pr}^{n-j}, \quad \text{for } n > 1. \end{aligned}$$

DEFINITION 4.21. The (rational valued) reduced Chern character is the series in the Chern classes

$$\bigoplus_{p>1} \frac{1}{p!} \mathrm{pr}^p.$$

The twisted reduced Chern character is the series

$$\mathrm{ch}_+ = \bigoplus_p \mathrm{ch}_p = \bigoplus_p \frac{(2\pi i)^p}{p!} \mathrm{pr}^p$$

In particular, we write

$$\mathrm{ch}_p = \frac{(2\pi i)^p}{p!} \mathrm{pr}^p \in H^{2p}\big(B.\mathrm{GL}(\mathbb{C}), \mathbb{Q}(p)\big)$$

for the component of degree $2p$ of the twisted reduced Chern character.

For other equivalent definitions of the Chern character and his properties the reader is referred to [**38**, §10].

4.5. The Stable Homotopy of the General Linear Group

We know from Section 3.4 that the Hurewicz morphism

$$\pi_j(\mathrm{GL}_n(\mathbb{C}), e) \to P_j(\mathrm{GL}_n(\mathbb{C}), \mathbb{Z}).$$

is an isomorphism after tensoring with $\mathbb{Q}$. The aim of this section is to describe the exact behaviour of the above morphisms when n goes to infinity. For proofs the reader is referred to [**22**] and to [**48**].

As in the case of the homology, we have

$$\pi_*(\mathrm{GL}(\mathbb{C}), e) = \lim_{n\to\infty} \pi_*(\mathrm{GL}_n(\mathbb{C}), e).$$

The structure of the homotopy groups of $\mathrm{GL}(\mathbb{C})$ is completely determined by the Bott Periodicity Theorem [**10**]. For any topological space X, let us denote by ΩX the loop space of X. As in the case of the group GL, we will denote by $\mathrm{SL}(\mathbb{C})$ the limit of the groups $\mathrm{SL}_n(\mathbb{C})$.

THEOREM 4.22 (Bott's Periodicity Theorem). *There is a weak homotopy equivalence* $h\colon |B.\,\mathrm{GL}(\mathbb{C})| \to \Omega\,\mathrm{SL}(\mathbb{C})$.

Since $\mathrm{GL}(\mathbb{C})$ is homeomorphic to $\mathrm{SL}(\mathbb{C}) \times \mathbb{C}^*$, we have that $\pi_j(\mathrm{GL}(\mathbb{C}), e) = \pi_{j+2}(\mathrm{GL}(\mathbb{C}), e)$, for $j \geq 0$. By induction this implies:

COROLLARY 4.23. *The homotopy groups of the infinite general linear group are given, for* $j \geq 0$, *by*

$$\pi_j(\mathrm{GL}(\mathbb{C}), e) = \begin{cases} 0, & \text{if } j \text{ is even,} \\ \mathbb{Z}, & \text{if } j \text{ is odd.} \end{cases}$$

Moreover Bott's Periodicity Theorem allows us to inductively determine the Hurewicz morphism (see [**22**]).

THEOREM 4.24. *Let* ϵ_{2p-1} *be a generator of the group* $\pi_{2p-1}(\mathrm{GL}(\mathbb{C}), e)$, *and let* Hur *be the Hurewicz morphism. Then*

$$\mathrm{Hur}(\epsilon_{2p-1}) = \pm(p-1)!\beta_{2p-1}.$$

REMARK 4.25. For each $p \geq 1$, the component of degree $2p$ of the twisted Chern character satisfies

$$\mathrm{ch}_p = \frac{(2\pi i)^p}{(p-1)!} b_p + \text{decomposable elements} \in H^{2p}\big(B.\mathrm{GL}(\mathbb{C}), \mathbb{Q}(p)\big).$$

Thus, if $\mathfrak{s}$ is the suspension, then $\mathfrak{s}(\mathrm{ch}_p) = (2\pi i)^p \alpha_{2p-1}/(p-1)!$. Therefore

$$\mathfrak{s}(\mathrm{ch}_p)\big(\mathrm{Hur}(\epsilon_{2p-1})\big) = \pm(2\pi i)^p.$$

4.6. Other Consequences of Bott's Periodicity Theorem

In the previous section we recalled a particular case of Bott's Periodicity Theorem for the unitary group, whose stable homotopy is periodic of period 2. But Bott's Periodicity Theorem ([**10**], see [**15, 48**]) is more general and establishes that the stable homotopy of other classical Lie groups and homogeneous spaces is periodic of period 8. Let us write $\mathrm{U} = \varinjlim \mathrm{U}_n$, $\mathrm{O} = \varinjlim \mathrm{O}_n(\mathbb{R})$ and $\mathrm{Sp} = \varinjlim \mathrm{Sp}_n$. Then SU and SO will have the obvious meaning. The inclusions $\mathrm{U}_n \subset \mathrm{O}_{2n}$, $\mathrm{Sp}_n \subset \mathrm{U}_{2n}$ and $\mathrm{O}_n \subset \mathrm{U}_n$ induce inclusions $\mathrm{U} \subset \mathrm{O}$, $\mathrm{Sp} \subset \mathrm{U}$ and $\mathrm{O} \subset \mathrm{U}$. Then the Bott's Periodicity Theorem imply

THEOREM 4.26. *The homotopy groups of the spaces* Sp, U / Sp, O / U, O, U / O *are periodic of period* 8. *Moreover, these groups are given by the following table*

X	0	1	2	3	4	5	6	7
Sp	0	0	0	$\mathbb{Z}$	$\mathbb{Z}/2\mathbb{Z}$	$\mathbb{Z}/2\mathbb{Z}$	0	$\mathbb{Z}$
U / Sp	0	$\mathbb{Z}$	0	0	0	$\mathbb{Z}$	$\mathbb{Z}/2\mathbb{Z}$	$\mathbb{Z}/2\mathbb{Z}$
O / U	$\mathbb{Z}/2\mathbb{Z}$	0	$\mathbb{Z}$	0	0	0	$\mathbb{Z}$	$\mathbb{Z}/2\mathbb{Z}$
O	$\mathbb{Z}/2\mathbb{Z}$	$\mathbb{Z}/2\mathbb{Z}$	0	$\mathbb{Z}$	0	0	0	$\mathbb{Z}$
U / O	0	$\mathbb{Z}$	$\mathbb{Z}/2\mathbb{Z}$	$\mathbb{Z}/2\mathbb{Z}$	0	$\mathbb{Z}$	0	0

where the groups listed in the ith column are the groups $\pi_j(X, x)$ *for* $j \equiv i \bmod 8$.

Unlike their finite dimensional counterpart, the infinite dimensional homogeneous spaces U / Sp, O / U and U / O have a structure of H-spaces ([**15**]). Therefore the above theorem and Cartan–Serre Theorem 3.17 allow us to compute the rank of the primitive part of its homology groups.

COROLLARY 4.27. *The dimension of the group* $P_m(\mathrm{U} / \mathrm{O}, \mathbb{R})$ *is one if* $m \equiv 1 \bmod 4$ *and zero otherwise.*

CHAPTER 5

Lie Algebra Cohomology and the Weil Algebra

5.1. de Rham Cohomology of a Lie Group

Let G be a real Lie group. Let $E^*(G,\mathbb{R})$ be the complex of global real valued differential forms on G. For any element $g \in G$, let l_g (resp. r_g) be the map given by the left action (resp. the right action) of g on G. That is $l_g(x) = gx$ and $r_g(x) = xg$. A differential form ω is called left invariant if $l_g^*\omega = \omega$ for all $g \in G$. The subspace of left invariant forms is a subcomplex of $E^*(G,\mathbb{R})$ denoted by $E^*(G,\mathbb{R})_{\mathrm{L}}$. We define right invariant forms analogously. The complex of right invariant forms will be denoted $E^*(G,\mathbb{R})_{\mathrm{R}}$. A differential form is called invariant if it is left and right invariant. We will denote by $E^*(G,\mathbb{R})_{\mathrm{I}}$ the subspace of invariant differential forms. Any invariant differential form is closed (see for instance [**32**, 4.9]).

The cohomology of the complex $E^*(G,\mathbb{R})$ is the de Rham cohomology of G. It is naturally isomorphic to the singular cohomology, $H^*(G,\mathbb{R})$. In the sequel we will identify both spaces. The cohomology of the complex $E^*(G,\mathbb{R})_{\mathrm{L}}$ is called left invariant cohomology and will be denoted by $H^*_{\mathrm{L}}(G,\mathbb{R})$. We have morphisms of algebras

$$E^*(G,\mathbb{R})_{\mathrm{I}} \to H^*_{\mathrm{L}}(G,\mathbb{R}) \to H^*(G,\mathbb{R}).$$

Let us denote by $\mathfrak{g}$ the Lie algebra of G, and let $\mathfrak{g}^\vee$ be the dual real vector space. In the exterior algebra $\Lambda^*\mathfrak{g}^\vee$ there is a unique derivation such that

$$d\colon \mathfrak{g}^\vee \to \mathfrak{g}^\vee \wedge \mathfrak{g}^\vee$$

is the dual of the Lie bracket. Explicitly this derivation is given by

$$d\omega(h_0,\dots,h_p) = \sum_{i<j}(-1)^{i+j}\omega([h_i,h_j],h_0,\dots,\hat{h}_i,\dots,\hat{h}_j,\dots,h_p).$$

Since d has square zero, $(\Lambda^*\mathfrak{g}^\vee, d)$ is a differential graded algebra. We will denote the differential graded algebra $(\Lambda^*\mathfrak{g}^\vee, d)$ by $E^*(\mathfrak{g},\mathbb{R})$, or by $E^*(\mathfrak{g})$ if there is no danger of confusion with the coefficients.

DEFINITION 5.1. The *cohomology algebra of the Lie algebra* $\mathfrak{g}$, denoted by $H^*(\mathfrak{g},\mathbb{R})$, is the cohomology of the graded differential algebra $E^*(\mathfrak{g})$.

The group G acts on itself by conjugation. This induces an action of G on $\mathfrak{g}$, called the adjoint action and denoted by Ad. The derivative of Ad is an action of $\mathfrak{g}$ over itself that we will denote by θ. This action is the

Lie bracket: $\theta(h)(x) = [h, x]$. Both actions, Ad and θ, can be extended to actions on $E^*(\mathfrak{g})$. The last one is given explicitly by

$$\theta(h)(\Phi)(h_1, \dots, h_p) = -\sum_{i=1}^{p} \Phi(h_1, \dots, [h, h_i], \dots, h_p). \tag{5.1}$$

Let us denote by $\big(E^*(\mathfrak{g})\big)^{\mathrm{Ad}}$ the subalgebra of invariant elements and let us write

$$\big(E^*(\mathfrak{g}, \mathbb{R})\big)_{\theta=0} = \{\Phi \in E^*(\mathfrak{g}, \mathbb{R}) \mid \theta(h)(\Phi) = 0, \forall h\}$$

Clearly there is an inclusion $\big(E^*(\mathfrak{g})\big)^{\mathrm{Ad}} \to \big(E^*(\mathfrak{g})\big)_{\theta=0}$. Moreover, if G is connected, both subspaces agree.

Evaluation at the unit element determines an isomorphism of differential graded algebras

$$E^*(G, \mathbb{R})_{\mathrm{L}} \to E^*(\mathfrak{g}).$$

This isomorphism induces an isomorphism

$$E^*(G, \mathbb{R})_{\mathrm{I}} \to \big(E^*(\mathfrak{g})\big)^{\mathrm{Ad}}.$$

Thus we obtain a commutative diagram, where the vertical arrows are isomorphisms.

$$\begin{array}{ccccc} E^*(G, \mathbb{R})_{\mathrm{I}} & \longrightarrow & H^*_{\mathrm{L}}(G, \mathbb{R}) & \longrightarrow & H^*(G, \mathbb{R}) \\ \downarrow & & \downarrow & & \\ \big(E^*(\mathfrak{g})\big)^{\mathrm{Ad}} & \longrightarrow & H^*(\mathfrak{g}). & & \end{array}$$

In general, the horizontal arrows of this diagram are not isomorphisms. But in the case of compact connected groups we have the following result:

PROPOSITION 5.2. *Let G be a compact connected Lie group. Then the inclusions*

$$i\colon E^*(G, \mathbb{R})_{\mathrm{L}} \to E^*(G, \mathbb{R}) \quad \text{and} \quad j\colon E^*(G, \mathbb{R})_{\mathrm{I}} \to E^*(G, \mathbb{R})$$

are homotopy equivalences. In particular the induced maps

$$E^*(\mathfrak{g})_{\theta=0} \to H^*(\mathfrak{g}, \mathbb{R}) \to H^*(G, \mathbb{R})$$

are isomorphisms.

SKETCH OF PROOF. Let dg be a normalized Haar measure on G, that is, a normalized invariant measure. Since G is compact, we can use this measure to average any differential form with respect to the left action of G over itself. Namely we define a morphism

$$\rho\colon E^*(G, \mathbb{R}) \to E^*(G, \mathbb{R})_{\mathrm{L}}$$

by

$$\rho(\omega) = \int_G g^*\omega \, dg.$$

Then ρ is an homotopy equivalence quasi-inverse of i (see for instance [**32**, 4.3]). To prove that j is a homotopy equivalence we can consider the action T of $G \times G$ on G given by

$$T_{a,b}(x) = axb^{-1}$$

and also use an averaging argument. □

From this result on compact real Lie groups we can obtain an analogous result for complex reductive groups. A *complex reductive group* is a complex analytic group that has a faithful analytical representation and such that every finite dimensional analytical representation is semisimple. Examples of reductive groups are the semisimple complex analytical groups and the general linear group $GL_n(\mathbb{C})$. The main structure theorem of complex reductive groups is the following (see [**39**, XV Theorem 3.1 and XVII Theorems 5.1 and 5.3])

THEOREM 5.3. *Let G be a complex analytic reductive Lie group. Then there is a real maximal compact Lie subgroup U of G such that G is isomorphic, as real analytical manifold, to the product $U \times E$, where E is a real vector space. Moreover, if $\mathfrak{g}$ is the complex Lie algebra of G and $\mathfrak{u}$ is the real Lie algebra of U, then*

$$\mathfrak{g} = \mathfrak{u} \otimes \mathbb{C}.$$

Let G be a closed analytic subgroup of $GL_n(\mathbb{C})$ that is connected and reductive. Let us denote by $\Omega^*(G)$ the complex of global holomorphic differential forms. Since G is a Stein manifold, the sheaves of holomorphic forms are acyclic. Therefore there is a natural isomorphism

$$H^*\big(\Omega^*(G)\big) \to H^*(G, \mathbb{C}).$$

Let $\mathfrak{g}$ denote the complex Lie algebra of G, and let us denote by $\mathfrak{g}^\vee$ the complex dual. As in the case of real Lie groups, we can define a complex graded differential algebra $E^*(\mathfrak{g}, \mathbb{C}) = \Lambda^*\mathfrak{g}^\vee$, and we can identify $E^*(\mathfrak{g}, \mathbb{C})$ with the subspace of left invariant holomorphic differential forms, $\Omega^*(G)_{\mathrm{L}}$.

COROLLARY 5.4. *The natural morphism*

$$H^*\big(E^*(\mathfrak{g}, \mathbb{C})\big) \to H^*\big(\Omega^*(G)\big)$$

is an isomorphism.

PROOF. Let U be a real maximal compact Lie subgroup of G. Let $\mathfrak{u}$ be its real Lie algebra. By Theorem 5.3 we have that

$$H^*(U, \mathbb{C}) = H^*(G, \mathbb{C}), \quad \text{and} \quad E^*(\mathfrak{g}, \mathbb{C}) = E^*(\mathfrak{u}, \mathbb{R}) \otimes \mathbb{C}.$$

Therefore the corollary is a direct consequence of Proposition 5.2. □

5.2. Reductive Lie Algebras

In this section we will review some results on the cohomology of reductive Lie algebras. The main references are [**11, 33**]. Let us fix a field of characteristic zero, k. Through this section Lie algebra will mean finite dimensional Lie algebra over k.

Let $\mathfrak{g}$ be a Lie algebra. A (finite dimensional) representation of $\mathfrak{g}$ is a Lie algebra homomorphism

$$\rho\colon \mathfrak{g} \to \mathfrak{gl}(V),$$

where V is a finite dimensional vector space.

DEFINITION 5.5. Let $\mathfrak{g}$ be a Lie algebra and ρ a representation. The *trace form* associated to ρ is the bilinear form

$$\langle X, Y\rangle_\rho = \operatorname{Tr}\bigl(\rho(X) \circ \rho(Y)\bigr).$$

The basic property of the trace form is:

$$\langle [X, Y], Z\rangle + \langle Y, [X, Z]\rangle = 0, \quad X, Y, Z \in \mathfrak{g}. \tag{5.2}$$

DEFINITION 5.6. A Lie algebra $\mathfrak{g}$ is called *reductive* if

$$\mathfrak{g} = Z_{\mathfrak{g}} \oplus \mathfrak{g}',$$

where $Z_{\mathfrak{g}}$ is the center of $\mathfrak{g}$ and $\mathfrak{g}'$ is a semisimple Lie algebra.

THEOREM 5.7. *Let $\mathfrak{g}$ be a finite dimensional Lie algebra. The following conditions are equivalent*

- $\mathfrak{g}$ *is reductive.*
- $\mathfrak{g}$ *admits a faithful, finite-dimensional representation with nondegenerate trace form.*
- $\mathfrak{g}$ *admits a faithful, finite-dimensional semisimple representation.*

EXAMPLE 5.8. If G is a compact Lie group then its Lie algebra is a real reductive Lie algebra. Analogously, if G is a complex reductive group then its Lie algebra is a complex reductive Lie algebra.

For any representation ρ of $\mathfrak{g}$ in a vector space V we will denote by $\rho(V)$ the subspace of V generated by the vectors $\rho(X)v$, for $X \in \mathfrak{g}$, and $v \in V$.

The adjoint action of $\mathfrak{g}$ on itself induces a representation, θ, of $\mathfrak{g}$ on the graded vector space $E^*(\mathfrak{g})$. Moreover this representation is compatible with the differential. Therefore we have induced representations in the subspace of cycles, denoted $Z^*(\mathfrak{g})$, the subspace of boundaries, denoted $B^*(\mathfrak{g})$ and the cohomology, $H^*(\mathfrak{g})$. The representation θ is semisimple, therefore there is a direct sum decomposition

$$E^*(\mathfrak{g}) = E^*(\mathfrak{g})_{\theta=0} \oplus \theta\bigl(E^*(\mathfrak{g})\bigr).$$

The relationship between invariant forms and Lie algebra cohomology is given by the following result.

LEMMA 5.9. *Let $\mathfrak{g}$ be a reductive Lie algebra. Then*

$$Z^*(\mathfrak{g}) = E^*(\mathfrak{g})_{\theta=0} \oplus B^*(\mathfrak{g}), \qquad B^*(\mathfrak{g}) = \theta\big(Z^*(\mathfrak{g})\big) = Z^*(\mathfrak{g}) \cap \theta\big(E^*(\mathfrak{g})\big).$$

As a direct consequence we have:

THEOREM 5.10. *Let $\mathfrak{g}$ be a reductive Lie algebra. Then the projection*

$$Z^*(\mathfrak{g}) \to H^*(\mathfrak{g})$$

induces an isomorphism

$$E^*(\mathfrak{g})_{\theta=0} \to H^*(\mathfrak{g}).$$

We can introduce in $E^*(\mathfrak{g})_{\theta=0}$ (hence in $H^*(\mathfrak{g})$) a structure of Hopf algebra. Let $\mu\colon \mathfrak{g} \oplus \mathfrak{g} \to \mathfrak{g}$ be the linear map given by $\mu(X, Y) = X + Y$. Then μ induces a linear morphism

$$\mu^*\colon E^*(\mathfrak{g}) \to E^*(\mathfrak{g} \oplus \mathfrak{g}).$$

Let

$$\eta\colon E^*(\mathfrak{g} \oplus \mathfrak{g}) \to E^*(\mathfrak{g})_{\theta=0} \otimes E^*(\mathfrak{g})_{\theta=0}$$

be the projection with kernel $\theta\big(E^*(\mathfrak{g} \oplus \mathfrak{g})\big)$. Let us write

$$\Delta = \eta \circ \mu^*|_{E^*(\mathfrak{g})}\colon E^*(\mathfrak{g})_{\theta=0} \to E^*(\mathfrak{g})_{\theta=0} \otimes E^*(\mathfrak{g})_{\theta=0}.$$

THEOREM 5.11. *The space $E^*(\mathfrak{g})_{\theta=0}$ provided with the wedge product, $\wedge$, the coproduct, Δ, and the unit and counit determined by the isomorphism $E^0(\mathfrak{g}) = k$, is a Hopf algebra.*

COROLLARY 5.12. *Let $P^*(\mathfrak{g})$ be the subspace of primitive elements of $E^*(\mathfrak{g})_{\theta=0}$. Then, the inclusion $P^*(\mathfrak{g}) \subset E^*(\mathfrak{g})_{\theta=0}$ induces an isomorphism $\bigwedge P^*(\mathfrak{g}) \cong E^*(\mathfrak{g})_{\theta=0}$.*

The structure of Hopf algebra and the subspace of primitive elements is functorial.

PROPOSITION 5.13. *Let $f\colon \mathfrak{h} \to \mathfrak{g}$ be a morphism of Lie algebras. Then the induced morphism*

$$f^*\colon E^*(\mathfrak{g}) \to E^*(\mathfrak{h}),$$

restricts to a morphism of Hopf algebras

$$f^*\colon E^*(\mathfrak{g})_{\theta=0} \to E^*(\mathfrak{h})_{\theta=0}.$$

In particular $f^*\big(P^*(\mathfrak{g})\big) = P^*(\mathfrak{h})$.

Let G be a compact Lie group, and let $\mathfrak{g}$ be its Lie algebra. The cohomology $H^*(G, \mathbb{R})$ has a structure of Hopf algebra induced by the multiplicative structure of G. On the other hand, by Proposition 5.2 there is a natural isomorphism $H^*(G, \mathbb{R}) \to H^*(\mathfrak{g}, \mathbb{R})$.

PROPOSITION 5.14. *Let G be a compact Lie group, and let $\mathfrak{g}$ be its Lie algebra. Then the natural isomorphism*

$$H^*(G, \mathbb{R}) \to H^*(\mathfrak{g}, \mathbb{R})$$

is a Hopf algebra isomorphism.

5.3. Characteristic Classes in de Rham Cohomology

In this section we will review the construction of characteristic classes in de Rham cohomology by means of a connection. We will work in the general case of a principal bundle. The basic references for this chapter are [**18, 24**].

Given a differentiable manifold M and a vector space V, we will denote by $E^*(M,V)$ the space of differential forms on M with values in V.

Let (r,E,π,B) be a G-principal bundle with base B and total space E. Since G acts on E on the right by r, we obtain a left action, r^* of G on $E^*(E,V)$.

Let $\mathfrak{g}$ be the Lie algebra of G and let $h \in \mathfrak{g}$. We can make h to act on $E^*(E,V)$ in two different ways. For each $x \in E$, the map $g \mapsto xg$ induces a morphism $\nu_x\colon \mathfrak{g} \to T_xE$. Let X_h be the fundamental vector field generated by h. This vector field is determined by the condition

$$(X_h)_x = \nu_x(h).$$

We will denote by $i(h)$ the substitution operator by the vector field X_h and by $\theta(h)$ the Lie derivative with respect to the vector field X_h. Explicitly, if $\Phi \in E^p(E,V)$,

$$\begin{aligned} i(h)\Phi(X_2,\dots,X_p) &= \Phi(X_h,X_2,\dots,X_p),\\ \theta(h)\Phi(X_1,\dots,X_p) &= X_h\Phi(X_1,\dots,X_p) - \sum_{i=1}^{p}\Phi(X_1,\dots,[X_h,X_i],\dots,X_p). \end{aligned}$$

The operators $i(h)$ and $\theta(h)$ are derivations (in the graded sense) of degree -1 and 0. They satisfy the following properties

$$\begin{aligned} i([h,k]) &= \theta(h)\circ i(k) - i(h)\circ\theta(k),\\ \theta([h,k]) &= \theta(h)\circ\theta(k) - \theta(h)\circ\theta(k),\\ \theta(h) &= i(h)\circ d + d\circ i(h),\\ d\circ\theta(h) &= \theta(h)\circ d. \end{aligned} \tag{5.3}$$

In particular, when the base is a point and $E = G$, we have operators $i(h)$ and $\theta(h)$ defined in $E^*(G,\mathbb{R})$. We may restrict these operators to the subalgebra $E^*(G,\mathbb{R})_{\mathrm{L}}$, that is, to $E^*(\mathfrak{g},\mathbb{R})$. In this case the operator $\theta(h)$ agrees with the operator of the same name defined in Section 5.1 ([**33**, 4.8]).

A differential form $\Phi \in E^*(E,V)$ is called horizontal if $i(h)\Phi = 0$ for all $h \in \mathfrak{g}$. The subset of horizontal differential forms is a subalgebra denoted $E^*(E,V)_{i=0}$. A differential form is called invariant if it is invariant under r^*. A differential form is called basic if it is both horizontal and invariant.

The induced morphism

$$\pi^*\colon E^*(B,V) \to E^*(E,V)$$

is injective and we may identify $E^*(B,V)$ with the subset of basic forms ([**32**, 6.3]). Moreover, if G is connected the subset of invariant forms agrees with the subset $E^*(E,V)_{\theta=0}$. For simplicity, in the sequel we will assume that G is connected.

The key concept in the theory of characteristic classes in de Rham cohomology is the concept of a connection.

DEFINITION 5.15. Let $E = (r, E, \pi, B)$ be a principal G-bundle. A *connection* in E is a 1-form $\nabla \in E^1(E, \mathfrak{g})$ such that

(1) $i(h)\nabla = h$.
(2) $r(g)^*\nabla = \mathrm{Ad}(g^{-1}) \circ \nabla$.

A connection ∇ induces a morphism

$$f_\nabla \colon E^1(\mathfrak{g}) \longrightarrow E^1(E, \mathbb{R}),$$

given by $f_\nabla(x) = x \circ \nabla$. We can extend this morphism multiplicatively to obtain a morphism of algebras.

$$f_\nabla \colon E^*(\mathfrak{g}) \longrightarrow E^*(E, \mathbb{R}).$$

EXAMPLE 5.16. Let us consider the principal bundle $G \to \{*\}$ with the usual right action $r_g g' = g'g$. Let l_g denote the usual left action. The *Maurer–Cartan connection* is the 1-form $\nabla^{\mathrm{MC}} \in E^1(G, \mathfrak{g})$ defined by

$$\nabla_g^{\mathrm{MC}} = (l_{g^{-1}})_*. \tag{5.4}$$

Then the morphism $f_{\nabla^{\mathrm{MC}}} \colon E^*(\mathfrak{g}) \to E^*(G, \mathbb{R})$ sends an element of $E^*(\mathfrak{g})$ to the corresponding left invariant form.

In general the morphism f_∇ is not a morphism of complexes. The map $\phi \colon E^1(\mathfrak{g}) \to E^2(E, \mathbb{R})$ given by $\phi(x) = df_\nabla(x) - f_\nabla(dx)$ measures how far is f_∇ from being a morphism of complexes. It is called the *curvature tensor* of the connection.

Let us denote by $S^*(\mathfrak{g}, \mathbb{R})$ or by $S^*(\mathfrak{g})$ the symmetric algebra over $\mathfrak{g}^\vee$. We view $S^*(\mathfrak{g})$ as a graded module by saying that the piece $S^p(\mathfrak{g})$ has degree $2p$.

DEFINITION 5.17. Let G be a Lie group and let $\mathfrak{g}$ be its Lie algebra. The *Weil algebra* of G is the bigraded algebra $W(G)$ defined by

$$W^{p,q}(G) = W^{p,q}(G, \mathbb{R}) = S^p(\mathfrak{g}) \otimes E^{q-p}(\mathfrak{g}).$$

Initially we will consider the Weil algebra only as a graded algebra with the total degree.

The map $f_\nabla \colon E^*(\mathfrak{g}) \to E^*(E, \mathbb{R})$ can be extended to a map defined on $W(G)$. To this end, for $x \in S^1(\mathfrak{g}) = \mathfrak{g}^\vee$, we write $f_\nabla(x) = \phi(x)$ and we extend f_∇ multiplicatively. By definition f_∇ is a morphism of algebras.

We want to define in $W(G)$ a differential, d, and operators $i(h)$ and $\theta(h)$, for $h \in \mathfrak{g}$, which are derivations of degree 1, -1 and 0 and that satisfy the conditions (5.3).

We first define the operator $i(h)$. It is already defined in $E^*(\mathfrak{g})$. We put $i(h) = 0$ in $S^*(\mathfrak{g})$. Thus there is a unique derivation $i(h)$ extending these data.

To define $\theta(h)$ in $W(G)$, it is enough to define it for $S^1(\mathfrak{g}) = \mathfrak{g}^\vee$. Here we may use the natural action θ of $\mathfrak{g}$ on $\mathfrak{g}^\vee$ (see formula (5.1)).

Now we want to introduce the differential in $W(G)$. Let us denote by d_E the differential in the complex $E^*(\mathfrak{g})$. And let

$$h\colon E^1(\mathfrak{g}) = \mathfrak{g}^\vee \to S^1(\mathfrak{g}) = \mathfrak{g}^\vee$$

be the identity map.

If $x \in E^1(\mathfrak{g})$ we write

$$dx = d_E x + h(x).$$

Let $x_1, \dots, x_p$ be a basis of $\mathfrak{g}$. Let $x'_1, \dots, x'_p$ be the dual basis. If $x \in S^1(\mathfrak{g})$, we write

$$dx = \sum_{k=1}^{p} \theta(x_k)x \otimes x'_k.$$

Then d can be extended uniquely to a derivation of degree one in $W(G)$.

For a proof of the following result see [**18**]

THEOREM 5.18. *Let (r, E, π, B) be a principal G-bundle and let ∇ be a connection. Then the map*

$$f_\nabla\colon W(G) \to E^*(E, \mathbb{R})$$

is a morphism of differential graded algebras compatible with the operators $\theta(h)$ and $i(h)$ for all $h \in \mathfrak{g}$.

Let $W^{2p}(G)_{i=0,\theta=0}$ be the subspace of basic elements of degree $2p$ of $W(G)$. It agrees with the subalgebra of invariant elements $S^p(\mathfrak{g})_{\theta=0}$. We will denote it by I_G^{2p}. The morphism $f_\nabla\colon W(G) \to E^*(E, \mathbb{R})$ sends I_G^* to the subalgebra of basic elements of $E^*(E, \mathbb{R})$ which has been identified with $E^*(B, \mathbb{R})$. Therefore, since the elements of I_G^* are closed we obtain a morphism of graded algebras

$$\omega_E\colon I_G^* \to H^*(B, \mathbb{R}).$$

DEFINITION 5.19. The *Chern-Weil morphism* is the morphism

$$\omega_E : I_G^* \to H^*(B, \mathbb{R}).$$

The next result is the heart of the de Rham realization of characteristic classes (cf. for instance [**17**])

THEOREM 5.20. *The Chern–Weil morphism is independent of the connection.*

As a consequence of this theorem, the image $\omega_E(I_G)$ is a subalgebra of $H^*(B, \mathbb{R})$ which is characteristic of the principal G-bundle E.

REMARK 5.21. Since the Weil algebra and the subspace of invariant elements only depend on the Lie algebra $\mathfrak{g}$ we will sometimes write $W(\mathfrak{g})$ and $I(\mathfrak{g})$ for $W(G)$ and I_G. Moreover, observe that we can define the Weil algebra and the subspace of invariant elements for a Lie algebra over any field.

We can think of the algebra $W(G)$ as an algebraic analogue of the de Rham algebra of the total space of the universal principal G-bundle. For instance, in the next section, we will see that the Weil algebra can be used to compute the suspension map. The first ingredient of this analogy is that the Weil algebra also has trivial cohomology. For a proof see [**17**, Theorem 1].

THEOREM 5.22. *The cohomology of* $W(G)$ *is*

$$H^{+}\big(W(G)\big) = 0 \quad \text{and} \quad H^{0}\big(W(G)\big) = \mathbb{R}.$$

To fully develop the relationship between $W(G)$ and $E^*(E.G, \mathbb{R})$ we have to introduce connections in simplicial principal G-bundles. We will use the complex of simplicial differential forms, $E^*_{\text{simp}}(E.G, \mathbb{R})$, because it is a commutative and associative algebra. Let $\pi \colon E. \to B.$ be a simplicial principal bundle. A *connection* for π is a differential form $\nabla \in E^1_{\text{simp}}(E., \mathfrak{g})$, such that its restriction to $H^p \times E_p$ is a connection for the principal bundle $H^p \times E_p \to H^p \times B_p$.

Let us choose a connection ∇ for the universal bundle $E.G \to B.G$. Since $E^*_{\text{simp}}(E.G, \mathbb{R})$ is a differential graded commutative and associative algebra, we have a morphism

$$f_\nabla \colon W(G) \to E^*_{\text{simp}}(E.G, \mathbb{R})$$

and a Chern–Weil morphism

$$\omega_{E.G} \colon I^*_G \to H^*(B.G, \mathbb{R}).$$

Observe that this Chern–Weil morphism is functorial on the group G. Moreover, if G is compact we have the following result (cf. [**24**]).

THEOREM 5.23. *Let G be a connected compact Lie group. Then the Chern–Weil morphism for the universal bundle*

$$\omega_{E.G} \colon I^*_G \to H^*(B.G, \mathbb{R})$$

is an isomorphism.

EXAMPLE 5.24. Let V be a complex vector space of dimension n. For any endomorphism $\phi \in \mathfrak{gl}(V)$ we will write $\wedge^p\phi$ the morphism induced in $\bigwedge^p V$. We define bilinear maps

$$\Box \colon \mathfrak{gl}(\textstyle\bigwedge^p V) \times \mathfrak{gl}(\bigwedge^q V) \to \mathfrak{gl}(\bigwedge^{p+q} V),$$

writing

$$(\phi \,\Box\, \psi)(x_1 \wedge \cdots \wedge x_{p+q}) = \frac{1}{p!q!} \sum_{\sigma \in \mathfrak{S}^{p+q}} (-1)^\sigma \phi(x_{\sigma(1)} \wedge \cdots \wedge x_{\sigma(p)}) \wedge \psi(x_{\sigma(p+1)} \wedge \cdots \wedge x_{\sigma(p+q)}).$$

In particular

$$\underbrace{\phi \,\Box \cdots \Box\, \phi}_{p} = p! \wedge^p \phi.$$

DEFINITION 5.25. Let V be an n-dimensional complex vector space. The *pth characteristic coefficient* is the element

$$C_p \in I^{2p}\big(\mathfrak{gl}(V)\big),$$

given by $C_0 = 1$, and

$$C_p(\phi_1, \dots, \phi_p) = \operatorname{Tr}(\phi_1 \,\square \cdots \square\, \phi_p).$$

Observe that the homogeneous polynomials

$$C_p(\phi) = \frac{1}{p!} C_p(\phi, \dots, \phi) = \operatorname{Tr}(\wedge^p \phi)$$

are the coefficients of the characteristic polynomial of the endomorphism ϕ.

DEFINITION 5.26. The *trace coefficients* of the complex vector space V are the elements $\operatorname{Tr}_p \in I^{2p}\big(\mathfrak{gl}(V)\big)$ defined by

$$\operatorname{Tr}_0 = \dim V$$

and

$$\operatorname{Tr}_p(\phi_1, \dots, \phi_p) = \sum_{\sigma \in \mathfrak{S}^p} \operatorname{Tr}\big(\phi_{\sigma(1)} \circ \cdots \circ \phi_{\sigma(p)}\big).$$

The trace coefficients and the characteristic coefficients are related by (see [**33**, A.3])

$$\operatorname{Tr}_1 = C_1,$$
$$\operatorname{Tr}_n = (-1)^{n+1} n C_n + \sum_{j=1}^{n-1} (-1)^{j+1} C_j \operatorname{Tr}_{n-j}, \quad \text{for } n > 1.$$

PROPOSITION 5.27. *Let* U_n *be the unitary group. Then the elements* C_p, $p = 0, \dots, n$ *generate* $I^*_{\mathrm{U}_n} \otimes \mathbb{C}$ *as an algebra. Moreover, the class* $\omega_{E.\,\mathrm{U}_n}(C_p) = c_p$ *agrees with the twisted Chern class of the universal rank* n *vector bundle. The elements* Tr_p, $p = 0, \dots, n$ *also generate* $I^*_{\mathrm{U}_n} \otimes \mathbb{C}$ *as an algebra, and the class* $\omega_{E.\,\mathrm{U}_n}(\operatorname{Tr}_p)/p!$ *agrees with* ch_p, *the component of degree* $2p$ *of the twisted Chern character of the universal rank* n *vector bundle.*

REMARK 5.28. Let G be a compact Lie group, and T be a maximal torus. Then $T = S^1 \times \cdots \times S^1$. Let $\mathfrak{w}$ be the Weyl group. Thus $I^*_G = (I^*_T)^{\mathfrak{w}}$ and $I^*_T = I^*_{S^1} \otimes \cdots \otimes I^*_{S^1}$. Therefore by functoriality and multiplicativity, the Chern–Weil morphism for any compact group G, $\omega_{E.G}$, is completely determined by the Chern-Weil morphism for the group S^1, $\omega_{E.S^1}$. This result is a version of the splitting principle.

The universal principal bundle has a canonical connection which allows us to obtain canonical representatives of the characteristic classes. This connection is defined using the Maurer–Cartan connection, ∇^{MC}, (see Example 5.16).

Let $q_i \colon H^p \times E_p G \to G$, $i = 0, \dots p$, denote the projection over the ith factor of $E_p G$. Let us write $\nabla_i = q_i^* \nabla^{\mathrm{MC}}$ and let

$$\nabla_{E.G} = t_0 \nabla_0 + \cdots + t_p \nabla_p, \tag{5.5}$$

where $t_0, \ldots, t_p$ are baricentric coordinates of H^p. Since $\sum t_i = 1$, the form

$$\nabla_{E.G} \in E^1_{\text{simp}}(E.G, \mathbb{R})$$

is a connection, and it is called the *canonical connection* of the universal principal bundle.

Let $\varphi : H \to G$ be a morphism of Lie groups. Then we have that

$$(E.\varphi)^* \nabla_{E.G} = \varphi_* \circ \nabla_{E.H}. \tag{5.6}$$

From the canonical connection we obtain a morphism

$$f_{\nabla_{E.G}} : I^*_G \to E^*_{\text{simp}}(B.G, \mathbb{R}).$$

A direct consequence of the equation (5.6) is that this morphism is functorial on the Lie group G.

REMARK 5.29. Composing with the integration morphism

$$E^*_{\text{simp}}(B.G, \mathbb{R}) \to s\mathcal{C}E^*(B.G, \mathbb{R})$$

we obtain also a well defined morphism

$$f_{\nabla_{E.G}} : I^*_G \to s\mathcal{C}E^*(B.G, \mathbb{R}),$$

which is also determined by the case $G = S^1$. It is easy to show that the image of this morphism is included in $\mathcal{N}E^*(B.G, \mathbb{R})$.

Changing the ground field from $\mathbb{R}$ to $\mathbb{C}$, the definition of the Weil algebra and the Weil homomorphism carries over to the case of complex analytical groups. Moreover, using Theorem 5.3, as in Corollary 5.4, we obtain the analogue of Theorem 5.23 for connected complex reductive groups:

THEOREM 5.30. *Let G be a complex connected analytic reductive Lie group. Then the complex valued Chern–Weil morphism for the universal bundle*

$$\omega_{E.G} \colon I^*_G \to H^*(B.G, \mathbb{C})$$

is an isomorphism.

5.4. The Suspension in the Weil Algebra

Recall that to define the suspension map in a principal G-bundle (r, E, π, B) we choose a point $x \in B$ and an inclusion $i \colon G \to E$ of G as the fibre over x. Let us start by describing the simplicial analogue of this situation.

The classifying space $B.G$ has $B_0G = \{e\}$. Therefore it is naturally a pointed space. Let $e.$ denote the simplicial point given by $e_n = (e, \ldots, e)$. The fibre of the universal principal G-bundle at this point is the simplicial manifold $G.$, which has $G_n = G$ and all faces and degeneracies equal to the identity. Let Δ denote the inclusion

$$\Delta \colon G. \to E.G$$

as the diagonal. There is a homotopy equivalence

$$\epsilon\colon E^*_{\text{simp}}(G.,\mathbb{R}) \to E^*(G,\mathbb{R}),$$

which sends a sequence of forms $\{\eta^{(n)}\}_n$ to the form $\eta^{(0)}$. The composition $\epsilon \circ \Delta^*$ sends a sequence of forms $\{\eta^{(n)}\}_n$, with

$$\eta^{(n)} \in E^*(H^n \times E_nG,\mathbb{R}),$$

to the form

$$\eta^{(0)} \in E^*(E_0G,\mathbb{R}) = E^*(G,\mathbb{R}).$$

The morphism $\epsilon \circ \Delta^*$ is the simplicial analogue of the morphism i^*.

Let us denote also by

$$\epsilon\colon W(G) \to \frac{W(G)}{S^+(\mathfrak{g}) \otimes E^*(\mathfrak{g})} = E^*(\mathfrak{g})$$

the natural projection. Then there is a commutative diagram

$$\begin{array}{ccccc} I^*_G & \longrightarrow & W(G) & \xrightarrow{\epsilon} & E^*(\mathfrak{g}) \\ \downarrow{\scriptstyle f_{\nabla_{E.G}}} & & \downarrow{\scriptstyle f_{\nabla_{E.G}}} & & \downarrow{\scriptstyle f_{\nabla^{\mathrm{MC}}}} \\ E^*_{\text{simp}}(B.G,\mathbb{R}) & \longrightarrow & E^*_{\text{simp}}(E.G,\mathbb{R}) & \xrightarrow{\epsilon\circ\Delta^*} & E^*(G,\mathbb{R}). \end{array}$$

If G is a compact connected Lie group, then the vertical arrows are quasi-isomorphisms. Thus, in this case we can compute the suspension map as follows (compare with Section 4.3). Let $\alpha \in I^+_G$. Since α is closed and $W(G)$ is acyclic, there exists an element β such that $d\beta = \alpha$. Then

$$\mathfrak{s}\big(\omega_{E.G}(\alpha)\big) = [\epsilon(\beta)]. \tag{5.7}$$

Let us give another description of the suspension morphism. We can give it in a purely algebraic setting. So let k a field of characteristic zero and let $\mathfrak{g}$ any Lie algebra over k.

Let us denote by F the filtration of the Weil algebra associated with the first degree.

$$F^pW(\mathfrak{g}) = \bigoplus_{p'\geq p} W^{p',*}(\mathfrak{g}).$$

We will call this filtration the *Hodge filtration.* By the definition of the Weil algebra, it is clear that

$$H^{2p}\big(F^pW(\mathfrak{g})\big) = I^{2p}(\mathfrak{g}).$$

Let us denote by

$$\delta\colon H^{2p-1}\big(E^*(\mathfrak{g})\big) = H^{2p-1}\big(W^*(\mathfrak{g})/F^1W^*(\mathfrak{g})\big) \to H^{2p}\big(F^1W^*(\mathfrak{g})\big),$$

the connecting morphism for the exact sequence

$$0 \to F^1W(\mathfrak{g}) \to W(\mathfrak{g}) \to W(\mathfrak{g})/F^1W(\mathfrak{g}) \to 0.$$

Since $W(\mathfrak{g})$ is acyclic, δ is an isomorphism.

DEFINITION 5.31. Let $\mathfrak{g}$ be a Lie algebra. Then the suspension $\mathfrak{s}_{\mathfrak{g}}$, is the composite map

$$\begin{aligned} I^{2p}(\mathfrak{g}) &= H^{2p}(F^p W(\mathfrak{g})) \\ &\longrightarrow H^{2p}(F^1 W(\mathfrak{g})) \\ &\xrightarrow{\delta^{-1}} H^{2p-1}(W(\mathfrak{g})/F^1 W(\mathfrak{g})) \\ &= H^{2p-1}(\mathfrak{g}). \end{aligned}$$

This definition is equivalent to the definition given at the beginning of the section. Thus the suspension just defined for the Lie algebra is compatible with the suspension in a compact Lie group or a complex reductive group. More concretely we have the following results.

PROPOSITION 5.32. *Let G be a compact connected Lie group, and let $\mathfrak{g}$ be its Lie algebra. Then the suspension can be factored as*

$$H^{2p}(B.G, \mathbb{R}) \xrightarrow{\omega_{E.G}^{-1}} I_G^{2p} \xrightarrow{\mathfrak{s}_{\mathfrak{g}}} H^{2p-1}(\mathfrak{g}) = H^{2p-1}(G, \mathbb{R}).$$

PROPOSITION 5.33. *Let G be a connected complex reductive Lie group, and let $\mathfrak{g}$ be its Lie algebra. Then the suspension can be factored as*

$$H^{2p}(B.G, \mathbb{C}) \xrightarrow{\omega_{E.G}^{-1}} I_G^{2p} \xrightarrow{\mathfrak{s}_{\mathfrak{g}}} H^{2p-1}(\mathfrak{g}) = H^{2p-1}(G, \mathbb{C}).$$

Moreover, the suspension computed using real linear algebra and the suspension computed using complex linear algebra are compatible.

PROPOSITION 5.34. *Let $\mathfrak{g}_{\mathbb{R}}$ be a real Lie algebra and let us write $\mathfrak{g} = \mathfrak{g}_{\mathbb{R}} \otimes \mathbb{C}$. Then there is a commutative diagram*

$$\begin{array}{ccc} I^{2p}(\mathfrak{g}_{\mathbb{R}}) & \longrightarrow & I^{2p}(\mathfrak{g}) \\ \downarrow{\scriptstyle \mathfrak{s}_{\mathfrak{g}_{\mathbb{R}}}} & & \downarrow{\scriptstyle \mathfrak{s}_{\mathfrak{g}}} \\ H^{2p-1}(\mathfrak{g}_{\mathbb{R}}, \mathbb{R}) & \longrightarrow & H^{2p-1}(\mathfrak{g}, \mathbb{C}). \end{array}$$

The relationship between the suspension and the subspaces of primitive and indecomposable elements of Theorem 4.18 has a complete parallelism in the case of Lie algebras (see [**33**, §6.14]).

THEOREM 5.35. *Let $\mathfrak{g}$ be a reductive Lie algebra. Then*

$$\begin{aligned} \operatorname{Im}(\mathfrak{s}_{\mathfrak{g}}) &= P^*(\mathfrak{g}), \\ \operatorname{Ker}(\mathfrak{s}_{\mathfrak{g}}) &= I^+(\mathfrak{g}) \cdot I^+(\mathfrak{g}). \end{aligned}$$

The suspension in the Lie algebra can be realized as an explicit linear morphism $I^{2p}(\mathfrak{g}) \mapsto E^{2p-1}(\mathfrak{g})_{\theta=0}$ (see [**33**, §6.8]).

PROPOSITION 5.36. *Let* $\Psi \in I^{2p}(\mathfrak{g})$. *Then* $\mathfrak{s}_{\mathfrak{g}}(\Psi)$ *is represented by the form*

$$\mathfrak{s}_{\mathfrak{g}}(\Psi)(x_1,\dots,x_p) = \frac{(-1)^{p-1}(p-1)!}{2^{p-1}(2p-1)!} \times \sum_{\sigma\in\mathfrak{S}^{2p-1}} (-1)^{\sigma}\Psi(x_{\sigma(1)},[x_{\sigma(2)},x_{\sigma(3)}],\dots,[x_{\sigma(2p-2)},x_{\sigma(2p-1)}]).$$

EXAMPLE 5.37. By Proposition 5.27, ch_p, the component of degree $2p$ of the twisted Chern character, is represented by $\mathrm{Tr}_p/p! \in I^{2p}_{\mathrm{U}_n}$, where the trace coefficients Tr_p is defined in Definition 5.26. Therefore $\mathfrak{s}(\mathrm{ch}_p)$ is represented, in $H^{2p-1}(\mathfrak{u}_n,\mathbb{C})$ by the form $\Phi_{2p-1} = \mathfrak{s}(\mathrm{Tr}_p)/p!$ given by

$$\Phi_{2p-1}(x_1,\dots,x_{2p-1}) = \frac{(-1)^{p-1}(p-1)!}{(2p-1)!} \sum_{\sigma\in\mathfrak{S}^{2p-1}} (-1)^{\sigma}\,\mathrm{Tr}(x_{\sigma(1)}\circ\dots\circ x_{\sigma(2p-1)}).$$

5.5. Relative Lie Algebra Cohomology

Let $\mathfrak{g}$ be a Lie algebra over a field of characteristic zero k, and let $\mathfrak{h}$ be a subalgebra. Let us denote by $\theta_{\mathfrak{h}}$ the restriction of the action θ to $\mathfrak{h}$. Since $\mathfrak{h}$ is a subalgebra, we have an induced action of $\mathfrak{h}$ on the quotient vector space $\mathfrak{g}/\mathfrak{h}$. Let us write

$$E^n(\mathfrak{g},\mathfrak{h}) = \mathrm{Hom}_{\mathfrak{h}}(\Lambda^n(\mathfrak{g}/\mathfrak{h}),k),$$

where we consider k as a trivial $\mathfrak{g}$ module. Then $E^n(\mathfrak{g},\mathfrak{h})$ is a subgroup of $E^n(\mathfrak{g})$. Moreover, since

$$d\big(E^n(\mathfrak{g},\mathfrak{h})\big) \subset E^{n+1}(\mathfrak{g},\mathfrak{h}),$$

we have that $E^*(\mathfrak{g},\mathfrak{h})$ is a subcomplex of $E^*(\mathfrak{g})$.

DEFINITION 5.38. The *relative Lie algebra cohomology groups* are the cohomology groups of the complex $E^*(\mathfrak{g},\mathfrak{h})$. These groups will be denoted by $H^*(\mathfrak{g},\mathfrak{h})$.

EXAMPLE 5.39. Let G be a connected Lie group and let H be a subgroup. Let us write $M = H\backslash G$. Then $E^*(\mathfrak{g},\mathfrak{h})$ is naturally isomorphic to the complex of G-invariant differential forms on M, $E^*(M,\mathbb{R})^G$. Moreover, if G is compact, an averaging argument as in Proposition 5.2 shows that there is an isomorphism

$$H^*(\mathfrak{g},\mathfrak{h}) \to H^*(M,\mathbb{R}).$$

DEFINITION 5.40. Let $\mathfrak{g}$ be a Lie algebra and let $j\colon \mathfrak{h}\to\mathfrak{g}$ be a subalgebra. We say that $\mathfrak{h}$ is *reductive in* $\mathfrak{g}$ if the representation $\theta_{\mathfrak{h}}$ is semisimple. We will say that $\mathfrak{h}$ is *noncohomologous to zero* if it is reductive in $\mathfrak{g}$ and the morphism

$$j^*\colon H^*(\mathfrak{g}) \to H^*(\mathfrak{h})$$

is surjective.

Our interest in the noncohomologous to zero subalgebras is the following result ([**33**, §10.18])

PROPOSITION 5.41. *Let* $\mathfrak{h}$ *be a noncohomologous to zero subalgebra of* $\mathfrak{g}$. *Then the morphism*

$$H^*(\mathfrak{g}, \mathfrak{h}) \to H^*(\mathfrak{g})$$

is injective.

CHAPTER 6

Group Cohomology and the van Est Isomorphism

In this section we will recall the notions of group cohomology and of continuous group cohomology. Moreover we will recall the van Est isomorphism that relates continuous cohomology with relative Lie algebra cohomology. We will follow [**35**]. For more details the reader may consult [**9**].

6.1. Group Homology and Cohomology

Let G be an abstract group. The homology and cohomology groups of G can be defined in many equivalent ways. One of them is the following. Let X be a contractible topological space, where G acts freely and properly discontinuously on the right; this means that the stabilizer of each point is trivial and that, for each point $x \in X$, there exists a neighbourhood U of x such that $xG \cap U = \{x\}$. Then the topological space X/G is a $K(G,1)$ space. That is, the fundamental group of X/G is G and the other homotopy groups are zero. For any abelian group A we put

$$H_*(G,A) = H_*(X/G,A) \quad \text{and} \quad H^*(G,A) = H^*(X/G,A).$$

In particular, the topological space $|E.G|$ is contractible and the right action of G on $E.G$ is free and properly discontinuous. Therefore

$$H_*(G,A) = H_*(|B.G|,A) \quad \text{and} \quad H^*(G,A) = H^*(|B.G|,A). \tag{6.1}$$

In order to distinguish between group homology and singular homology, if G is a topological group, we will denote $H^*_{\text{group}}(G,A)$ and $H_*^{\text{group}}(G,A)$ for the cohomology and homology of G as an abstract group. This distinction will not be necessary when G is a discrete group or an abstract group. Thus, in this case we will denote the group cohomology groups and the group homology groups by $H^*(G,A)$ and $H_*(G,A)$ respectively.

Let us recall another way to define the cohomology groups of G. A G-module is an abelian group E provided with a left action of G by automorphisms. A G-morphism is a morphism of abelian groups compatible with the action of G. An injective G-morphism is called *strong* if it has a left inverse as a morphism of abelian groups. A G-morphism $f\colon E \to F$ is called *strong* if the injective G-morphisms $\operatorname{Ker} f \to E$ and $E/\operatorname{Ker} f \to F$ are strong.

The group cohomology groups are derived functors in the framework of relative homological algebra. This means that we will consider mainly strong morphisms. This amounts to considering only exact sequences which are split as sequences of abelian groups. Examples of this point of view are the notions of strong resolution and of relatively injective G-modules.

DEFINITION 6.1. Let E be a G-module. A *strong resolution* of E is a long exact sequence

$$0 \to E \to E^0 \to E^1 \to \cdots$$

such that all morphisms are strong morphisms.

DEFINITION 6.2. A G-module I is called *relatively injective* if, for any strong injective G-morphism $f\colon E \to F$ and any G-morphism $u\colon E \to I$, there exists a G-morphism $v\colon F \to I$ such that $v \circ f = u$.

DEFINITION 6.3. Let Γ_G be the functor from the category of G-modules to the category of abelian groups, that sends any G-module E to its subgroup of invariant elements $\Gamma_G(E) = E^G$. The functor Γ_G is left exact. The *group cohomology groups* of a G-module E are the relative right derived functors of Γ_G

$$H^i(G, E) = R^i\Gamma_G(E).$$

The adjective relative in this definition means that, to compute the cohomology groups of a G-module E, we need a strong resolution

$$0 \to E \to E^0 \to E^1 \to \cdots,$$

with all the G-modules E^i being relatively injective. Then the cohomology groups of E are the cohomology groups of the complex $(E^*)^G$.

Let A be an abelian group. We will consider A as a G module with the trivial action. Let us denote by $\mathcal{F}(E_nG, A)$ the group of maps from E_nG to A. Then $\mathcal{F}(E.G, A)$ is a cosimplicial abelian group and let $\mathcal{CF}(E.G, A)$ be the associated complex. Since the group G acts on $E.G$ on the right, there is an induced left action on $\mathcal{CF}(E.G, A)$. Observe that

$$\mathcal{F}(E_nG, A)^G = \mathcal{F}(B_nG, A).$$

For a proof of the following result see [**35**].

PROPOSITION 6.4. *The G-modules $\mathcal{F}(E_nG, A)$ are relatively injective and the complex*

$$0 \to A \to \mathcal{CF}(E_0G, A) \to \mathcal{CF}(E_1G, A) \to \cdots$$

is a strong resolution. The same is true if we replace the complex $\mathcal{CF}(E.G, A)$ by the normalization $\mathcal{NF}(E.G, A)$.

As a consequence of Proposition 6.4 we recover equation (6.1):

$$H^*(G, A) = H^*(|B.G|, A).$$

As in the case of singular homology, there is a stability result for the homology of the series of classical groups. In particular we will use the following result established in [**41, 59**].

THEOREM 6.5. *Let A be a local ring. Let us write*

$$\mathrm{GL}(A) = \varinjlim \mathrm{GL}_n(A).$$

Then the natural morphism

$$H_k(\mathrm{GL}_n(A), \mathbb{Z}) \to H_k(\mathrm{GL}(A), \mathbb{Z})$$

is an isomorphism for $k \leq (n-1)/2$.

6.2. Continuous Group Cohomology

Let now G be a connected Lie group. The theory of group cohomology carries over to the category of continuous (or differentiable) G-modules and we obtain the theory of *continuous group cohomology.* See [**35**, Chapter III] for the details. The continuous cohomology groups of a continuous G-module E will be denoted by $H^*_{\mathrm{cont}}(G, E)$. We are interested in the continuous G-module $\mathbb{R}$ with the usual topology and the trivial G action. Let $E^0(E_nG, \mathbb{R})$ denote the space of differentiable functions. The differentiable analogue of Proposition 6.4 is:

PROPOSITION 6.6. *The complexes*

$$\mathcal{C}E^0(E.G, \mathbb{R}) \quad \textit{and} \quad \mathcal{N}E^0(E.G, \mathbb{R})$$

are strong resolutions of the continuous G-module $\mathbb{R}$ by relatively injective continuous G-modules. In consequence the continuous cohomology groups of $\mathbb{R}$ are the cohomology groups of either complex

$$\mathcal{C}E^0(B.G, \mathbb{R}) \quad \textit{or} \quad \mathcal{N}E^0(B.G, \mathbb{R}).$$

Let us construct another resolution of the G-module $\mathbb{R}$. Let U be a maximal compact subgroup of G. Let M be the homogeneous space $M = U\backslash G$. Then M is diffeomorphic to a real vector space. Thus it is contractible ([**39**, XV, Theorem 3.1]). There is a right action of G on M defined, for $p = Ux$, by $r_g(p) = Uxg$. This right action induces a left action of G on the complex of differential forms $E^*(M, \mathbb{R})$.

PROPOSITION 6.7. *The complex*

$$0 \to \mathbb{R} \to E^0(M, \mathbb{R}) \to E^1(M, \mathbb{R}) \to \cdots$$

is a strong resolution by relatively injective G-modules. Therefore the continuous cohomology groups of the G-module $\mathbb{R}$ are the cohomology groups of the complex $E^*(M, \mathbb{R})^G$.

By this proposition and Example 5.39 we can relate continuous cohomology and relative Lie algebra cohomology.

PROPOSITION 6.8. *Let G be a connected Lie group, U a maximal compact subgroup and $M = U\backslash G$. Let $\mathfrak{g}$ and $\mathfrak{u}$ be the Lie algebras of G and U. The isomorphism $E^*(G, \mathbb{R})_R \to E^*(\mathfrak{g}, \mathbb{R})$ induces an isomorphism*

$$E^*(M, \mathbb{R})^G \to E^*(\mathfrak{g}, \mathfrak{u}, \mathbb{R}).$$

From this proposition we obtain the *van Est isomorphism*:

THEOREM 6.9 (van Est). *With the hypotheses of the above proposition, there is a natural isomorphism*

$$H^*_{\mathrm{cont}}(G, \mathbb{R}) \to H^*(\mathfrak{g}, \mathfrak{u}, \mathbb{R}).$$

6.3. Computation of Continuous Cohomology

The van Est isomorphism can be used to compute the continuous cohomology of Lie groups. Let G be an algebraic reductive group defined over $\mathbb{R}$. Let us assume that $G(\mathbb{R})$ is a connected Lie group. Then $G(\mathbb{C})$ is a complex connected reductive group. Let $\mathfrak{g}_{\mathbb{R}}$ be the Lie algebra of $G(\mathbb{R})$. Then $\mathfrak{g}_{\mathbb{C}} = \mathfrak{g} \otimes \mathbb{C}$ is the Lie algebra of $G(\mathbb{C})$.

Let K be a maximal compact subgroup of $G(\mathbb{R})$, let $\mathfrak{k}$ be the Lie algebra of K and let

$$\mathfrak{g}_{\mathbb{R}} = \mathfrak{k} \oplus \mathfrak{p}$$

be the Cartan decomposition of $\mathfrak{g}_{\mathbb{R}}$ with respect to $\mathfrak{k}$. Then

$$\mathfrak{g}_u = \mathfrak{k} \oplus i\mathfrak{p} \subseteq \mathfrak{g}_{\mathbb{C}}$$

is a compact form of $\mathfrak{g}_{\mathbb{R}}$. Thus, the corresponding Lie group G_u is a maximal compact subgroup of $G(\mathbb{C})$ that contains K.

Let us write $X = K\backslash G(\mathbb{R})$ and $X_u = K\backslash G_u$. Strictly speaking, X_u is the compact twin of the symmetric space X and is determined by the pair $(G(\mathbb{R}), K)$. But since it is determined up to isomorphism by the group $G(\mathbb{R})$, and in all the applications the compact subgroup K will be fixed, we will denote X_u by $\mathrm{CT}\big(G(\mathbb{R})\big)$ and call it the compact twin of $G(\mathbb{R})$. We consider the right actions of $G(\mathbb{R})$ on X and of G_u on X_u given by $r_g(p) := pg$. As in Proposition 6.8, the complex $E^*(\mathfrak{g}_{\mathbb{R}}, \mathfrak{k}, \mathbb{R})$ is isomorphic to the complex of $G(\mathbb{R})$-invariant forms on X and $E^*(\mathfrak{g}_u, \mathfrak{k}, \mathbb{R})$ to the complex of G_u-invariant forms on X_u. Since X_u is compact, an averaging argument as in Proposition 5.2, shows that the last isomorphism induces an isomorphism

$$\alpha\colon H^*(X_u, \mathbb{R}) \to H^*(\mathfrak{g}_u, \mathfrak{k}, \mathbb{R}).$$

Let now

$$\iota\colon E^*(\mathfrak{g}_u, \mathfrak{k}, \mathbb{R}) \to E^*(\mathfrak{g}_{\mathbb{R}}, \mathfrak{k}, \mathbb{R})$$

be the isomorphism that sends $\omega \in E^j(\mathfrak{g}_u, \mathfrak{k}, \mathbb{R}) = \mathrm{Hom}_{\mathfrak{k}}(\bigwedge^j i\mathfrak{p}, \mathbb{R})$ to $\iota(\omega) = i^j\omega \in E^j(\mathfrak{g}_{\mathbb{R}}, \mathfrak{k}, \mathbb{R}) = \mathrm{Hom}_{\mathfrak{k}}(\bigwedge^j \mathfrak{p}, \mathbb{R})$. Let us denote by γ' the composition

$$H^*(X_u, \mathbb{R}) \xrightarrow{\alpha} H^*(\mathfrak{g}_u, \mathfrak{k}, \mathbb{R}) \xrightarrow{\iota} H^*(\mathfrak{g}_{\mathbb{R}}, \mathfrak{k}, \mathbb{R}) \xrightarrow{\beta} H^*_{\mathrm{cont}}(G(\mathbb{R}), \mathbb{R}), \tag{6.2}$$

where β is the inverse of the van Est isomorphism.

Since α, β and ι are isomorphisms, γ' is an isomorphism. Therefore we can use the fact that the cohomology groups of the classical Lie groups and their associated homogeneous spaces are well known, to compute the continuous cohomology groups of the classical Lie groups.

EXAMPLE 6.10. Let us consider the group

$$G = \operatorname{Res}_{\mathbb{R}/\mathbb{C}} \mathrm{GL}_{n,\mathbb{C}}\,.$$

That is, $G(\mathbb{R})$ is the Lie group $\mathrm{GL}_n(\mathbb{C})$ but viewed as a real Lie group. Then the maximal compact subgroup of $G(\mathbb{R})$ is U_n. The complexification $G(\mathbb{C})$ is $\mathrm{GL}_n(\mathbb{C}) \times \mathrm{GL}_n(\mathbb{C})$, with the inclusion

$$\begin{aligned}\mathrm{GL}_n(\mathbb{C}) &\to \mathrm{GL}_n(\mathbb{C}) \times \mathrm{GL}_n(\mathbb{C})\\ M &\mapsto (\overline{M}, M)\end{aligned}$$

The complex conjugation $\tau\colon \mathrm{GL}_n(\mathbb{C}) \times \mathrm{GL}_n(\mathbb{C}) \to \mathrm{GL}_n(\mathbb{C}) \times \mathrm{GL}_n(\mathbb{C})$ is given by $\tau(M, N) = (\overline{N}, \overline{M})$. Therefore the compact subgroup of the complexification is $\mathrm{U}_n \times \mathrm{U}_n$ and the compact twin of $G(\mathbb{R})$ is homeomorphic to U_n.

Thus, in this case, $\mathfrak{g}_{\mathbb{R}}$ is $\mathfrak{gl}_n(\mathbb{C})$, but viewed as a real vector space. The Lie algebra of the compact subgroup is $\mathfrak{k} = \mathfrak{u}_n$. The subspace $\mathfrak{p}$ is the subspace of Hermitian matrices. The other Lie algebras are $\mathfrak{g}_{\mathbb{C}} = \mathfrak{gl}_n(\mathbb{C}) \oplus \mathfrak{gl}_n(\mathbb{C})$ and $\mathfrak{g}_u = \mathfrak{u}_n \oplus \mathfrak{u}_n$. With these identifications, the inclusions $\mathfrak{g}_{\mathbb{R}} \to \mathfrak{g}_{\mathbb{C}}$ and $\mathfrak{k} \to \mathfrak{g}_u$ are both given by $M \mapsto (\overline{M}, M)$. Therefore $\mathfrak{g}_u/\mathfrak{k} \cong \mathfrak{u}_n$, with the projection $(M, N) \mapsto N - \overline{M}$. We have that $E^*(\mathfrak{g}_u, \mathfrak{k}, \mathbb{R})$ is naturally isomorphic to $E^*(\mathfrak{u}_n, \mathbb{R})_{\theta=0}$.

PROPOSITION 6.11. *The continuous cohomology groups of* $\mathrm{GL}_n(\mathbb{C})$ *are*

$$H^*_{\mathrm{cont}}(\mathrm{GL}_n(\mathbb{C}), \mathbb{R}) \cong H^*(\mathrm{U}_n, \mathbb{R}) = \bigwedge(\alpha_1, \alpha_3, \dots, \alpha_{2n-1}).$$

By the naturality of the construction and the stability of the singular cohomology of the groups U_n, we obtain:

COROLLARY 6.12. *Let $n \leq n'$ and $m < 2n$ be positive integers. Then the morphism*

$$\varphi^*_{n',n}\colon H^m_{\mathrm{cont}}(\mathrm{GL}_{n'}(\mathbb{C}), \mathbb{R}) \to H^m_{\mathrm{cont}}(\mathrm{GL}_n(\mathbb{C}), \mathbb{R})$$

is an isomorphism.

CHAPTER 7

Small Cosimplicial Algebras

In this section we will recall briefly the notion of cosimplicial algebras and we will review the theory of small cosimplicial algebras and small differential graded algebras. This theory was introduced by Beilinson in order to compare his regulator with Borel's regulator. We will follow [**55**].

7.1. Cosimplicial Algebras

Let k be a field of characteristic zero, and let $\mathbf{A}$ be a k-linear tensorial category with unit $\mathbb{1}$, such that $\operatorname{End}(\mathbb{1}) = k$. During this section the category $\mathbf{A}$ will be fixed, and all objects will belong to this category. For simplicity, we will work as if $\mathbf{A}$ were a category of vector spaces, that is, we will work with elements.

The fact that $\mathbf{A}$ is a k-linear tensorial category with unity implies that the category of cosimplicial objects of A, $\mathcal{CS}(\mathbf{A})$, is also a k-linear tensorial category with unit. Explicitly, the tensor product in the category $\mathcal{CS}(\mathbf{A})$ is given by

$$(X \otimes Y)^n = X^n \otimes Y^n,$$

with

$$\tau(x \otimes y) = \tau(x) \otimes \tau(y),$$

for any increasing map $\tau\colon [n] \to [m]$. The unit element, that we will also denote by $\mathbb{1}$ is the constant cosimplicial object

$$\mathbb{1}^p = \mathbb{1},$$

with all the structure morphisms equal to the identity.

Let $X, Y \in \mathcal{CS}(\mathbf{A})$. The relationship between the tensor product of simplicial objects and the tensor product of complexes is given by the *Alexander–Whitney morphism* and the shuffle morphism (see for instance [**21**, Chapter VI, §12] or [**43**, Chapter VIII, §8]):

$$\mathcal{C}X \otimes \mathcal{C}Y \xrightarrow{\text{AW}} \mathcal{C}(X \otimes Y),$$

$$\mathcal{C}(X \otimes Y) \xrightarrow{S} \mathcal{C}X \otimes \mathcal{C}Y.$$

The morphism AW is given componentwise, for each pair of integers p, q, with $p + q = n$, by

$$\begin{aligned} X^p \otimes Y^q &\to X^n \otimes Y^n \\ x \otimes y &\mapsto \delta^n \circ \cdots \circ \delta^{p+1}(x) \otimes \delta^0 \circ \cdots \circ \delta^0(y). \end{aligned}$$

Let us explain the definition of the shuffle morphism. Let p, q be non negative integers. A (p, q) shuffle, (μ, ν), is a pair of disjoint sets of integers

$$1 \leq \mu_1 < \mu_2 < \cdots < \mu_p \leq p+q, \qquad 1 \leq \nu_1 < \nu_2 < \cdots < \nu_q \leq p+q.$$

The sign of the shuffle, $\epsilon(\mu, \nu)$, is the sign of the permutation

$$(\mu_1, \ldots, \mu_p, \nu_1, \ldots, \nu_q).$$

If $p+q=n$, the shuffle map is given componentwise by

$$S\colon X^n \otimes Y^n \longrightarrow X^p \otimes Y^q$$

$$x \otimes y \mapsto \sum_{(\mu,\nu)} \epsilon(\mu, \nu) \sigma^{\nu_1 - 1} \cdots \sigma^{\nu_q - 1} x \otimes \sigma^{\mu_1 - 1} \cdots \sigma^{\mu_p - 1} y,$$

where the sum runs over all shuffles.

Both the Alexander–Whitney morphism and the shuffle morphism respect the normalized complexes. Thus they induce morphisms

$$\mathcal{N}X \otimes \mathcal{N}Y \xrightarrow{\text{AW}} \mathcal{N}(X \otimes Y),$$

$$\mathcal{N}(X \otimes Y) \xrightarrow{S} \mathcal{N}X \otimes \mathcal{N}Y.$$

Moreover, at the level of normalized complexes they satisfy $S \circ \text{AW} = \text{Id}$ (see [**26**, Theorem 2.1a]).

DEFINITION 7.1. An (*associative*) *cosimplicial algebra* (*with unit*) in **A**, is an associative and unitary algebra object of the category $\mathcal{CS}(\mathbf{A})$. That is, it consists of an object $X \in \text{Ob}(\mathcal{CS}(\mathbf{A}))$, together with morphisms

$$\mathbb{1} \xrightarrow{e} X$$

$$X \otimes X \xrightarrow{\mu} X$$

satisfying the axioms of an associative algebra (see Section 3.1).

In this section all algebras will be associative and unitary, thus we will use the word algebra to refer to associative unitary algebra.

It is easy to see that the above definition is equivalent to saying that a cosimplicial algebra is a cosimplicial object X, such that each X^n, $n \geq 0$, is an algebra, and each structure morphism is a morphism of algebras.

The Alexander–Whitney morphism and the shuffle morphism allow us to transport the multiplicative structure of a cosimplicial algebra to its associated complex, and the multiplicative structure of a differential graded algebra (DGA for short) to its associated cosimplicial object.

Let X be a cosimplicial (associative) algebra. Then the morphisms

$$\mathbb{1} = \mathcal{C}\mathbb{1} \xrightarrow{\mathcal{C}e} \mathcal{C}X,$$

$$\cup\colon \mathcal{C}X \otimes \mathcal{C}X \xrightarrow{\text{AW}} \mathcal{C}(X \otimes X) \xrightarrow{\mathcal{C}(\mu)} \mathcal{C}X,$$

provide $\mathcal{C}X$ with a structure of DGA. Explicitly, if $x \in X^p$, $y \in X^q$, then

$$x \cup y = \delta^{p+q} \circ \cdots \circ \delta^{p+1}(x) \cdot \delta^0 \circ \cdots \circ \delta^0(y).$$

By the properties of the Alexander–Whitney morphism, the associativity of X implies the associativity of $\mathcal{C}X$. On the other hand, if X is commutative, then $\mathcal{C}X$ does not need to be (graded) commutative, but only commutative up to homotopy.

Let us see that the complex of degenerate elements, $\mathcal{D}X$, is a left ideal of $\mathcal{C}X$. Let $x \in X^p$ and $y \in X^{q-1}$. Then

$$\begin{aligned} x \cup \delta^i(y) &= \delta^{p+q} \circ \cdots \circ \delta^{p+1}(x) \cdot (\delta^0)^p \circ \delta^i(y) \\ &= \delta^{i+p} \circ \delta^{p+q-1} \circ \cdots \circ \delta^{p+1}(x) \cdot \delta^{i+p} \circ (\delta^0)^p(y) \\ &= \delta^{i+p}(x \cup y). \end{aligned} \tag{7.1}$$

Analogously one can see that $\mathcal{D}X$ is also a right ideal. Moreover, we have that $\mathcal{N}X$ is a subalgebra: let $x \in \mathcal{N}X^p$ and $y \in \mathcal{N}X^q$. Then, if $i \leq p-1$,

$$\begin{aligned} \sigma^i(x \cup y) &= \sigma^i(\delta^{p+q} \circ \cdots \circ \delta^{p+1}(x) \cdot (\delta^0)^p(y)) \\ &= \sigma^i(\delta^{p+q} \circ \cdots \circ \delta^{p+1}(x)) \cdot \sigma^i((\delta^0)^p(y)) \\ &= \delta^{p+q-1} \circ \cdots \circ \delta^p \circ \sigma^i(x) \cdot (\delta^0)^{p-1}(y) \\ &= 0, \end{aligned}$$

whereas, if $i \geq p$,

$$\begin{aligned} \sigma^i(x \cup y) &= \sigma^i(\delta^{p+q} \circ \cdots \circ \delta^{p+1}(x) \cdot (\delta^0)^p(y)) \\ &= \sigma^i(\delta^{p+q} \circ \cdots \circ \delta^{p+1}(x)) \cdot \sigma^i((\delta^0)^p(y)) \\ &= \delta^{p+q-1} \circ \cdots \circ \delta^{p+1}(x) \cdot (\delta^0)^p \circ \sigma^i(y) \\ &= 0. \end{aligned}$$

As in the case of $\mathcal{C}X$, the commutativity of X does not imply the commutativity of $\mathcal{N}X$.

Let now Y be a DGA. We want to induce in $\mathcal{K}Y$ a structure of cosimplicial algebra. Since $\mathcal{K}\mathbb{1}$ is the constant cosimplicial unit object, which we have denoted also by $\mathbb{1}$, we can take as the unit of $\mathcal{K}Y$ the morphism

$$\mathcal{K}\mathbb{1} \xrightarrow{\mathcal{K}e} \mathcal{K}Y.$$

Let $\mu \colon Y \otimes Y \to Y$ be the product of Y. Let us consider the composition

$$\mathcal{N}(\mathcal{K}Y \otimes \mathcal{K}Y) \xrightarrow{S} \mathcal{N}\mathcal{K}Y \otimes \mathcal{N}\mathcal{K}Y = Y \otimes Y \xrightarrow{\mu} Y = \mathcal{N}\mathcal{K}Y.$$

Since $\mathcal{N}$ is an equivalence of categories, we obtain a morphism

$$\mathcal{K}Y \otimes \mathcal{K}Y \to \mathcal{K}Y.$$

The associativity of Y implies the associativity of $\mathcal{K}Y$. But in contrast with the Alexander–Whitney morphism, by the properties of the shuffle morphism (see for instance [**21**, Chapter VI, §12]) the commutativity of Y implies the commutativity of $\mathcal{K}Y$.

7.2. Small Algebras

The fact that, if X is a commutative cosimplicial algebra, $\mathcal{N}X$ does not need to be commutative, leads to the following definition.

DEFINITION 7.2 (Beilinson).
(1) A cosimplicial associative algebra X is called *small* if it is commutative, it is generated by X^0 and X^1 under the cup-product and, if $I = \mathcal{N}X^1 = \operatorname{Ker}\sigma^0 \subset X^1$ then $I^2 = 0$.
(2) A differential graded associative algebra Y is called *small* if it is commutative and it is generated, as an algebra, by Y^0 and Y^1.

For simplicity we will consider only reduced cosimplicial algebras and reduced DGA. A *reduced* cosimplicial algebra is a cosimplicial algebra X, such that $X^0 = \mathbb{1}$. Therefore $\delta^0 = \delta^1 \colon X^0 \to X^1$ are the unit morphism, and, by Corollary 2.10, there is a canonical decomposition

$$X^1 = \mathbb{1} \oplus \mathcal{N}X^1. \tag{7.2}$$

A *reduced* DGA is a DGA Y with $Y^0 = \mathbb{1}$. Therefore the differential $d^0 = 0$.

The main result about small algebras is the following one.

THEOREM 7.3. *The functors $\mathcal{N}$ and $\mathcal{K}$ induce an equivalence of categories between the category of reduced small cosimplicial algebras and the category of small differential graded algebras.*

PROOF. For the proof of this result we will follow [**55**].

LEMMA 7.4. *Let X be a reduced commutative cosimplicial algebra. Then X is generated (in the sense of the cup-product) by X^1 if and only if $\mathcal{N}X$ is generated by $\mathcal{N}X^1$.*

PROOF. Let us denote by m the morphism

$$\begin{aligned} m\colon X^1 \otimes \cdots \otimes X^1 &\to X^p \\ x_1 \otimes \cdots \otimes x_p &\mapsto x_1 \cup \cdots \cup x_p. \end{aligned}$$

The decomposition (7.2), $X^1 = \mathbb{1} \oplus \mathcal{N}X^1$, induces a decomposition of $(X^1)^{\otimes p}$. A typical summand being $\mathcal{N}X^1 \otimes \ldots \mathbb{1} \cdots \otimes \mathcal{N}X^1$, with the factor $\mathbb{1}$ in the positions $i_1 < \cdots < i_r$. By the same argument as in (7.1), the restriction of m to one of such summands is given by

$$m(x_1 \otimes \cdots 1 \cdots \otimes x_p) = \delta^{i_r - 1} \circ \cdots \circ \delta^{i_1 - 1}(x_1 \cup \cdots \cup x_p). \tag{7.3}$$

Therefore m sends the summand $\mathcal{N}X^1 \otimes \cdots \otimes \mathcal{N}X^1$ to $\mathcal{N}X^p$ and the other summands to $\mathcal{D}X^p$. Hence, if X is generated by X^1, then $\mathcal{N}X$ is generated by $\mathcal{N}X^1$. Assume now that $\mathcal{N}X$ is generated by $\mathcal{N}X^1$. Since $\mathcal{D}X$ is generated, as an abelian group, by elements of the form $\delta^{i_r} \circ \cdots \circ \delta^{i_1}(x)$, with $x \in \mathcal{N}X$, equation (7.3) implies that X is generated by X^1. □

LEMMA 7.5.
(1) *If X is a reduced small cosimplicial algebra, then $\mathcal{N}X$ is a reduced small DGA.*

(2) *If Y is a reduced small DGA, then $\mathcal{K}Y$ is a reduced small cosimplicial algebra, and the natural isomorphism $Y \cong \mathcal{N}\mathcal{K}Y$ is an isomorphism of algebras.*

PROOF. (1) Clearly $\mathcal{N}X$ is reduced, and by Lemma 7.4 it is generated by $\mathcal{N}X^1$. It remains to show the commutativity. Since $\mathcal{N}X^1$ generates $\mathcal{N}X$, it suffices to show that $x \cup x = 0$ for $x \in \mathcal{N}X^1$. Since

$$dx = \delta^0 x - \delta^1 x + \delta^2 x \in \mathcal{N}X^2,$$

and using again the fact that $\mathcal{N}X^1$ generates $\mathcal{N}X$ we have

$$\delta^1 x = \delta^0 x + \delta^2 x + \sum_i \alpha_i \cup \beta_i = \delta^0 x + \delta^2 x + \sum_i \delta^2 \alpha_i \delta^0 \beta_i,$$

for some elements $\alpha_i, \beta_i \in \mathcal{N}X^1$. Using the fact that X^1 is a commutative algebra, that $\mathcal{N}X^1$ is a square zero ideal and that the faces δ^j are morphisms of algebras we obtain

$$(\delta^1 x)^2 = \delta^2 x \delta^0 x + \delta^0 x \delta^2 x = 2x \cup x.$$

Therefore

$$2x \cup x = (\delta^1 x)^2 = \delta^1 x^2 = 0.$$

(2) That $\mathcal{K}Y$ is reduced is obvious. That it is commutative follows from the properties of the shuffle morphism. That $\mathcal{K}Y$ is generated by $\mathcal{K}Y^1$ follows from Lemma 7.4. Let us show that $\mathcal{N}\mathcal{K}Y^1$ is a square zero ideal. Let $x, y \in \mathcal{N}\mathcal{K}Y^1$. Then $x \otimes y \in \mathcal{N}(\mathcal{K}Y \otimes \mathcal{K}Y)$. Therefore

$$x \cdot y = \mu(S(x \otimes y)) = \mu(\sigma^0 x \otimes y + x \otimes \sigma^0 y) = 0,$$

where μ is the product in Y. Hence $\mathcal{K}Y$ is a small cosimplicial algebra. To prove that the multiplicative structures of Y and $\mathcal{N}\mathcal{K}Y$ agree, let us consider the following commutative diagram

$$\begin{array}{ccc}
Y \otimes Y & \longrightarrow & Y \\
\Big\| & & \Big\| \\
\mathcal{N}\mathcal{K}Y \otimes \mathcal{N}\mathcal{K}Y & \longrightarrow & \mathcal{N}\mathcal{K}Y \\
{\scriptstyle \mathrm{AW}}\Big\downarrow & & \Big\| \\
\mathcal{N}(\mathcal{K}Y \otimes \mathcal{K}Y) & \longrightarrow & \mathcal{N}\mathcal{K}Y \\
{\scriptstyle S}\Big\downarrow & & \Big\| \\
Y \otimes Y & \longrightarrow & Y,
\end{array}$$

where the fourth horizontal arrow defines the multiplicative structure of Y, and the first arrow is induced by the second one. Thus the first arrow corresponds to the multiplicative structure of $\mathcal{N}\mathcal{K}Y$. Since at the level of normalized cochains $S \circ \mathrm{AW} = \mathrm{Id}$ we obtain that both multiplicative structures agree. □

To prove the theorem it only remains to show that the functor $\mathcal{K}$ is essentially surjective. This will be a consequence of the next result.

LEMMA 7.6. *Let X be an object of $\mathcal{CS}(\mathbf{A})$, provided with two structures of small reduced cosimplicial algebras, (e, μ) and (e', μ'). If these structures induce the same multiplicative structure on $\mathcal{N}X$ then they agree.*

PROOF. That the unit morphisms e and e' agree is obvious. Since X is small and reduced for both algebra structures, in the decomposition $X^1 = \mathbb{1} \oplus \mathcal{N}X^1$, the summand $\mathcal{N}X^1$ is an ideal of square zero for both structures. Therefore μ and μ' agree on X^1. Let us denote by $\cup$ and $\cup'$ the cup-product induced in $\mathcal{C}X$ by μ and μ' respectively. Let us show that they agree. To this end, we only have to show that the two linear maps

$$m, m' \colon X^1 \otimes \cdots \otimes X^1 \to X^p,$$

induced by $\cup$ and $\cup'$, agree. This follows from equation (7.3) and the fact that $\cup$ and $\cup'$ agree on $\mathcal{N}X$.

Since the morphism m is surjective, to show that μ and μ' agree, we have to show that the morphisms $\mu \circ (m \otimes m)$ and $\mu' \circ (m \otimes m)$ agree. Since μ and μ' agree on X^1, this is a consequence of the bimultiplicativity of the cup-product, which holds by the commutativity of either algebra structure. □

Let now X be a small cosimplicial algebra, and let $\psi \colon X \to \mathcal{K}\mathcal{N}X$ be the isomorphism provided by the Dold–Kan correspondence. Since $\mathcal{N}(\psi)$ is an isomorphism of algebras, by Lemma 7.6, ψ is also an isomorphism of algebras. □

From now on we will focus our attention on two tensor categories.

(1) The category $\mathbf{Vec}_k$ of vector spaces over k. In this case the concepts of cosimplicial algebra and differential graded algebra are the usual ones. We will call them c-algebras and d-algebras respectively.

(2) The category $\mathcal{C}^+(\mathbf{Vec}_k)$ of cochain complexes of vector spaces in non-negative degrees. In this case we will call a cosimplicial algebra a cd-algebra and a DGA a dd-algebra. Thus a cd-algebra is a cosimplicial differential graded algebra. By convention we will assume that the differential degree is the first one and that the cosimplicial degree is the second one. The concepts of reduced and small will refer to the cosimplicial degree. A dd-algebra is a bigraded algebra with two differentials, d' and d'', of degree $(1, 0)$ and $(0, 1)$ respectively, with $d'd'' = d''d'$ and which are graded derivations with respect to the first and the second degree respectively. To be consistent with the corresponding concepts in the cosimplicial case, the concepts of small and reduced dd-algebras are not symmetrical with respect to the two degrees. A small dd-algebra will be a commutative dd-algebra, $X^{*,\cdot}$ generated by $X^{*,0}$ and $X^{*,1}$. The dd-algebra $X^{*,\cdot}$ will be reduced if $X^{0,0} = k$ and $X^{p,0} = 0$, for $p \geq 1$.

The functor {d-algebras} $\to$ {k-algebras} that send a DGA to its zeroth component has a left adjoint: the algebra of differentials Ω^* that sends a k-algebra R to the DGA $\Omega^*(R) = \Omega^*_{R/k}$. Therefore $\Omega^*(R)$ solves the following universal problem. Given a k-algebra R, a DGA X^* and a morphism of algebras $f\colon R \to X^0$, there is a unique morphism of DGAs $\Omega^*(R) \to X^*$ extending f.

Analogously we have functors

$$\{\text{cd-algebras}\} \to \{\text{c-algebras}\},$$
$$\{\text{dd-algebras}\} \to \{\text{d-algebras}\}$$

whose left adjoint is also denoted by Ω^*. Finally we may consider the functors

$$\{\text{reduced small cd-algebras}\} \to \{\text{reduced small c-algebras}\},$$
$$\{\text{reduced small dd-algebras}\} \to \{\text{reduced small d-algebras}\}.$$

They also have a left adjoint which we will denote by $\overline{\Omega}^*$. If X is a reduced small d-algebra then $\Omega^*(X)$ is reduced and small. Therefore in this case $\overline{\Omega}^* = \Omega^*$. On the other hand, if X is a reduced small c-algebra, then $\Omega^*(X)$ is reduced but not necessarily small. In this case

$$\overline{\Omega}^*(X) = \Omega^*(X)/\{\text{cd-ideal spanned by } [\operatorname{Ker}\sigma^0\colon \Omega^*(X^1) \to \Omega^*(X^0)]^2\}.$$

The fact that $\mathcal{N}$ is an equivalence of categories, between the category of small reduced c-algebras and the category of small reduced d-algebras, and also between the category of small reduced cd-algebras and the category of small reduced dd-algebras, implies the following result (see [**2**, Apendix A] and [**55**, §2]).

PROPOSITION 7.7. *Let X be a small reduced c-algebra. Then*

$$\overline{\Omega}(\mathcal{N}X) = \mathcal{N}\overline{\Omega}(X).$$

CHAPTER 8

Higher Diagonals and Differential Forms

8.1. The Sheaf of Differential Forms

Let X be a scheme over a field k. Let us denote by $\Delta\colon X \to X \times X$ the diagonal morphism, and by $\mathcal{I}_\Delta$ the sheaf of ideals of the subscheme $\Delta(X)$. Then there is a canonical isomorphism ([**37**, II.8])

$$\Omega^1_X \to \Delta^{-1}(\mathcal{I}_\Delta/\mathcal{I}_\Delta^2). \tag{8.1}$$

For instance, in the affine case, if $X = \operatorname{Spec}(A)$, the diagonal morphism is induced by the morphism of algebras

$$\begin{aligned} \mu\colon A \otimes A &\to A \\ f \otimes g &\mapsto fg. \end{aligned}$$

Therefore the ideal of the diagonal is $\operatorname{Ker}(\mu)$. The isomorphism (8.1) is given by the morphism

$$\begin{aligned} \varphi\colon \Omega^1_{A/k} &\to \operatorname{Ker}(\mu)/\operatorname{Ker}(\mu)^2 \\ df &\mapsto 1 \otimes f - f \otimes 1. \end{aligned}$$

The inverse is induced by the morphism

$$\begin{aligned} \psi\colon A \otimes A &\to \Omega^1_{A/k} \\ f \otimes g &\mapsto f\,dg. \end{aligned}$$

The aim of this section is to generalize this description of Ω^1_X to the sheaves Ω^n_X.

DEFINITION 8.1. Let $X.$ be a simplicial scheme. A *sheaf of ideals* $\mathcal{J}^{\cdot} \subset \mathcal{O}_{X.}$ is a family of sheaves of ideals $\{\mathcal{J}^n\}_n$, with $\mathcal{J}^n$ a sheaf of ideals on X_n, such that

$$\sigma_i^*(\mathcal{J}^n) \subset \mathcal{J}^{n-1}, \quad i = 0, \dots, n-1,$$

and

$$\delta_i^*(\mathcal{J}^n) \subset \mathcal{J}^{n+1}, \quad i = 0, \dots, n+1.$$

Let X be a scheme over a field k. Let $E.X$ be the simplicial scheme defined in Example 2.5. Observe that the degeneracy maps $\sigma_i\colon E_nX \to E_{n+1}X$ are given by

$$\sigma_i = \operatorname{Id}^i \times \Delta \times \operatorname{Id}^{n-i}, \quad i = 0, \dots, n,$$

and the face maps $\delta_i\colon E_nX \to E_{n-1}X$ by

$$\delta_i = \operatorname{Id}^i \times \varepsilon \times \operatorname{Id}^{n-i}, \quad i = 0, \dots, n,$$

where $\varepsilon\colon X \to \operatorname{Spec}(k)$ is the structural morphism.

Let us write $\mathcal{I}_\Delta = \operatorname{Ker}(\sigma_0^*) \subset \mathcal{O}_{E_1X} = \mathcal{O}_{X\times X}$. Let $\mathcal{J}^{\cdot}_\Delta$ be the ideal generated by $\mathcal{I}_\Delta^2$.

For each pair of integers $0 \le i < j \le n$ let us denote by

$$\delta_{\widehat{ij}}\colon E_nX \to E_1X$$

the morphism

$$\delta_{\widehat{ij}} = \varepsilon^i \times \operatorname{Id} \times \varepsilon^{j-i-1} \times \operatorname{Id} \times \varepsilon^{n-j}.$$

LEMMA 8.2. *The sheaf of ideals $\mathcal{J}^{\cdot}_\Delta$ is given by*

$$\mathcal{J}^n_\Delta = \sum_{\substack{i,j=0\\ i<j}}^{n} \delta^*_{\widehat{ij}}\mathcal{I}^2_\Delta$$

PROOF. Let us write $\mathcal{J}^n = \sum \delta^*_{\widehat{ij}}\mathcal{I}^2_\Delta$. The ideal $\mathcal{J}^{\cdot}_\Delta$ is the smallest ideal that contains $\mathcal{I}^2_\Delta$. Thus it is closed under the face and degeneracy morphisms. Therefore it contains $\mathcal{J}^{\cdot}$. On the other hand, using the commutation rules between faces and degeneracies and the fact that $\mathcal{I}_\Delta = \operatorname{Ker}\sigma_0^*$, it is easy to see that $\mathcal{J}^{\cdot}$ is closed under the face and the degeneracy morphisms. Therefore, since it is an ideal that contains $\mathcal{I}^2_\Delta$, it also contains $\mathcal{J}^{\cdot}_\Delta$. □

Let us denote by $E^{(1)}_{\cdot}X$ the simplicial subscheme of $E_{\cdot}X$ defined by the ideal $\mathcal{J}_\Delta$. Let $\Delta_n\colon X \to X^{n+1}$ be the diagonal morphism, $\Delta_n(x) = (x,\dots,x)$. The following result is a direct consequence of Lemma 8.2.

PROPOSITION 8.3. *The reduced scheme $(E^{(1)}_{\cdot}X)_{\mathrm{red}}$ is the diagonal subscheme $\Delta_{\cdot}(X)$.*

Since the simplicial scheme $\Delta_{\cdot}(X)$ is the constant simplicial scheme with $\Delta_n(X) = X$, and all the faces and degeneracies equal to the identity, we may think of $\mathcal{O}_{E^{(1)}_{\cdot}X}$ as a sheaf of cosimplicial algebras over X. Let us give another description of this sheaf.

The family $\Delta_{\cdot}^{-1}\mathcal{O}_{E_{\cdot}X} = \{\Delta_n^{-1}\mathcal{O}_{E_nX}\}_{n\in\mathbb{N}}$ is a cosimplicial sheaf of algebras over X. Explicitly we have

$$\Delta_n^{-1}\mathcal{O}_{E_nX} = \overbrace{\mathcal{O}_X \underset{k}{\otimes} \cdots \underset{k}{\otimes} \mathcal{O}_X}^{n+1},$$

$$\sigma^i(f_0\otimes\cdots\otimes f_n) = f_0\otimes\cdots\otimes f_{i-1}\otimes f_if_{i+1}\otimes f_{i+2}\otimes\cdots\otimes f_n,$$

$$\delta^i(f_0\otimes\cdots\otimes f_n) = f_0\otimes\cdots\otimes f_{i-1}\otimes 1\otimes f_i\otimes\cdots\otimes f_n.$$

The sheaf $\Delta_{\cdot}^{-1}\mathcal{J}^{\cdot}_\Delta$ is a cosimplicial sheaf of ideals of $\Delta_{\cdot}^{-1}\mathcal{O}_{E_{\cdot}X}$. Then

$$\mathcal{O}_{E^{(1)}_{\cdot}X} = \Delta_{\cdot}^{-1}\mathcal{O}_{E_{\cdot}X}/\Delta_{\cdot}^{-1}\mathcal{J}^{\cdot}_\Delta.$$

By construction $\mathcal{O}_{E^{(1)}_{\cdot}X}$ is a sheaf of small cosimplicial associative and commutative algebras. Although $\mathcal{O}_{E^{(1)}_{\cdot}X}$ is not reduced, the proof of

Lemma 7.5 can be applied to show that $\mathcal{N}(\mathcal{O}_{E^{(1)}_{\cdot}X})$ is a sheaf of differential graded commutative associative algebras.

THEOREM 8.4. *Let X be a regular scheme over a field k. The sequence of morphisms*

$$\begin{aligned} \psi_0^n\colon \qquad & \Delta_n^{-1}\mathcal{O}_{E_nX} \to \Omega^n_{X/k} \\ & f_0 \otimes f_1 \otimes \cdots \otimes f_n \mapsto f_0\, df_1 \wedge \cdots \wedge df_n \end{aligned}$$

induces a natural morphism of sheaves of differential algebras

$$\psi\colon \mathcal{N}(\Delta_{\cdot}^{-1}\mathcal{O}_{E.X}) \to \Omega^{\cdot}_{X/k},$$

and an isomorphism of sheaves of differential algebras

$$\overline{\psi}\colon \mathcal{N}(\mathcal{O}_{E^{(1)}_{\cdot}X}) \cong \Omega^{\cdot}_{X/k}.$$

PROOF. Since the morphisms of sheaves ψ^n are defined globally, to prove that they are zero on $\mathcal{J}^{\cdot}_{\Delta}$, and that they induce the desired isomorphism, is a local question. Thus we may assume that $X = \operatorname{Spec}(A)$, with A a local regular k-algebra. Let us denote by $A^{\cdot}$ the cosimplicial ring of sections of $\Delta_{\cdot}^{-1}\mathcal{O}_{E.X}$, by $J^{\cdot}_{\Delta}$ the cosimplicial ideal of sections of $\mathcal{J}^{\cdot}_{\Delta}$, by I_{Δ} the ideal of sections of $\mathcal{I}_{\Delta}$, by N^* the complex of sections of $\mathcal{N}\Delta_{\cdot}^{-1}\mathcal{O}_{E.X}$ and by $\overline{N}^*$ the complex of sections of $\mathcal{N}(\Delta_{\cdot}^{-1}\mathcal{O}_{E.X}/\Delta_{\cdot}^{-1}\mathcal{J}^{\cdot}_{\Delta})$. An easy computation shows that $\psi^n(J^n_{\Delta}) = 0$. Thus we have well defined morphisms $\overline{\psi}^n\colon \overline{N}^n \to \Omega^n_{A/k}$.

By definition,

$$\overline{N}^n = \bigcap_{i=0}^{n-1} \operatorname{Ker} \overline{\sigma}^i,$$

where $\overline{\sigma}^i\colon A^n/J^n_{\Delta} \to A^{n-1}/J^{n-1}_{\Delta}$ is the morphism induced by σ^i. Let us give a more pleasant presentation of $\overline{N}^n$.

LEMMA 8.5. *Let $A^{\cdot}$ be a cosimplicial ring and let $J^{\cdot}$ be a cosimplicial ideal. Then*

$$\bigcap_{i=0}^{n-1} (\sigma^i)^{-1}(J^{n-1}) = \bigcap_{i=0}^{n-1} \operatorname{Ker}\sigma^i + J^n.$$

PROOF. The fact that the right term is included in the left term is clear. To see the inclusion in the opposite direction, let

$$x \in \bigcap (\sigma^i)^{-1}(J^{n-1}).$$

Let us define inductively

$$x_0 = x - \delta_0\sigma_0 x, \quad x_i = x_{i-1} - \delta_i\sigma_i x_{i-1}.$$

Then, using the commutation rules between faces and degeneracies, and the fact that J is a cosimplicial ideal, we obtain that

$$x_{n-1} \in \bigcap_{i=0}^{n-1} \operatorname{Ker}\sigma^i \quad \text{and} \quad x - x_{n-1} \in J^n. \qquad \square$$

This lemma implies that

$$\overline{N}^n = \bigcap_{i=0}^{n-1} \operatorname{Ker} \overline{\sigma}^i = \frac{\bigcap_{i=0}^{n-1} \operatorname{Ker} \sigma^i}{\bigcap_{i=0}^{n-1} \operatorname{Ker} \sigma^i \cap J^n} = \frac{N^n}{N^n \cap J^n_\Delta}.$$

Using the regularity of A, and the fact that the ideals $\operatorname{Ker} \sigma^i$ correspond to regular subschemes that intersect properly, it follows that $\bigcap \operatorname{Ker} \sigma^i = \prod \operatorname{Ker} \sigma^i$. Thus

$$\overline{N}^n = \frac{\prod_{i=0}^{n-1} \operatorname{Ker} \sigma^i}{\prod_{i=0}^{n-1} \operatorname{Ker} \sigma^i \cap J^n_\Delta}.$$

Let us write $B = A \underset{k}{\otimes} A$. Then there is a natural isomorphism

$$\overbrace{B \underset{A}{\otimes} \cdots \underset{A}{\otimes} B}^{n} \cong \overbrace{A \underset{k}{\otimes} \cdots \underset{k}{\otimes} A}^{n+1}.$$

Observe that I_Δ is an ideal of B and that

$$\operatorname{Ker} \sigma^i = B \otimes \cdots \otimes \underset{i+1}{I_\Delta} \otimes \cdots \otimes B.$$

Therefore

$$N^n = \prod_{i=0}^{n-1} \operatorname{Ker} \sigma^i = \overbrace{I_\Delta \underset{A}{\otimes} \ldots \underset{A}{\otimes} I_\Delta}^{n}.$$

The following result follows from a direct computation.

LEMMA 8.6. *The restriction of ψ^n to N^n can be factored as*

$$N^n = I_\Delta^{\otimes n} \xrightarrow{\overbrace{\psi_1 \otimes \cdots \otimes \psi_1}^{n}} (\Omega^1_{A/k})^{\otimes n} \to \Omega^n_{A/k}.$$

Now we want to identify $J^n_\Delta \cap N^n$. First observe that the ideals of the form

$$I_\Delta \otimes \cdots \otimes \underset{i}{I^2_\Delta} \otimes \cdots \otimes I_\Delta$$

are contained in J^n_Δ. Therefore $\overline{N}^n$ can be identified with a quotient of $(I_\Delta / I^2_\Delta)^{\otimes n}$. In addition we have the following result.

LEMMA 8.7. *Let $x_1, \ldots, x_n$ be elements of I_Δ, and let $1 \le i < n$. Then*

$$x_1 \underset{A}{\otimes} \cdots \underset{A}{\otimes} (x_i \underset{A}{\otimes} x_{i+1} + x_{i+1} \underset{A}{\otimes} x_i) \underset{A}{\otimes} \cdots \underset{A}{\otimes} x_n \in J^n_\Delta.$$

PROOF. It is enough to treat the case $n = 2$, $i = 1$. Since $x_1, x_2 \in I_\Delta$, we can write

$$x_1 = \sum_i f_i \underset{k}{\otimes} g_i \quad \text{and} \quad x_2 = \sum_j s_j \underset{k}{\otimes} t_j,$$

with

$$\sum_i f_i g_i = \sum_j s_j t_j = 0.$$

But then

$$\begin{aligned}
&\sum_{i,j}(f_i \underset{k}{\otimes} g_i) \underset{A}{\otimes} (s_j \underset{k}{\otimes} t_j) + \sum_{i,j}(s_j \underset{k}{\otimes} t_j) \underset{A}{\otimes} (f_i \underset{k}{\otimes} g_i) \\
&\quad = \sum_{i,j} f_i \underset{k}{\otimes} g_i s_j \underset{k}{\otimes} t_j + s_j \underset{k}{\otimes} t_j f_i \underset{k}{\otimes} g_i \\
&\quad = \sum_{ij}(f_i \otimes 1 \otimes 1 - 1 \otimes f_i \otimes 1)(1 \otimes s_j \otimes 1 - s_j \otimes 1 \otimes 1)(1 \otimes g_i \otimes t_j) \\
&\qquad + \sum_{ij}(f_i \otimes 1 \otimes g_i)(s_j \otimes 1 \otimes t_j) + \sum_{ij}(f_i \otimes g_i \otimes 1)(s_j \otimes t_j \otimes 1) \\
&\qquad + \sum_{ij}(1 \otimes f_i \otimes 1 - f_i \otimes 1 \otimes 1)(s_j \otimes t_j \otimes 1)(1 \otimes 1 \otimes g_i).
\end{aligned}$$

Since all the terms on the right hand side belong to J_Δ^2 we have proved the lemma. □

As a consequence of this lemma, we have that $\overline{N}^n$ is a quotient of $\bigwedge^n(I_\Delta/I_\Delta^2)$. But since the composition

$$\bigwedge^n(I_\Delta/I_\Delta^2) \to \overline{N}^n \xrightarrow{\overline{\psi}^n} \Omega^n_{A/k}$$

is an isomorphism, we obtain that $\overline{\psi^n}$ is an isomorphism.

The fact that ψ is compatible with the differential is immediate.

Let us end the proof of the theorem discussing the compatibility with the product. Let us consider $A^{\cdot}$ a graded algebra with the cup-product. Then

$$A^n = \overbrace{A \underset{k}{\otimes} \cdots \underset{k}{\otimes} A}^{n+1}$$

and

$$f_0 \otimes \cdots \otimes f_n \cup g_0 \otimes \cdots \otimes g_m = f_0 \otimes \cdots \otimes f_n g_0 \otimes \cdots \otimes g_m.$$

Thus it is clear that $\psi^{\cdot}$ is not a morphism of algebras. On the other hand, if we consider the presentation

$$A^n = \overbrace{B \underset{A}{\otimes} \ldots \underset{A}{\otimes} B}^{n},$$

then, if $n, m > 0$,

$$f_1 \otimes \cdots \otimes f_n \cup g_1 \otimes \cdots \otimes g_m = f_1 \otimes \cdots \otimes f_n \otimes g_1 \otimes \cdots \otimes g_m.$$

Therefore, the fact that ψ restricted to $N^{\cdot}$ is a morphism of algebras follows from Lemma 8.6. □

8.2. The Weil Algebra Revisited

In this section we will apply Theorem 8.4 to the case of a complex algebraic reductive group and show that we can obtain the Chern–Weil morphism from the morphism $\overline{\psi}$.

Let G be a connected complex reductive group. Let $\mathcal{I}_e$ be the ideal of $\mathcal{O}_{B_1G} = \mathcal{O}_G$ defined by the unit point e. Let us denote by $\mathcal{J}_e$ the cosimplicial ideal generated by $\mathcal{I}_e^2$. Let $B_{\cdot}^{(1)}G$ be the subscheme defined by $\mathcal{J}_e$. The right action of G over $E_{\cdot}G$ induces a left action on the algebra of global sections $\mathcal{O}(E_{\cdot}G)$. The ideal $\mathcal{J}_\Delta(E_{\cdot}G)$ is invariant for this action. Therefore, there is an induced action of G in $\mathcal{O}(E_{\cdot}^{(1)}G)$. We have that

$$\mathcal{O}(E_{\cdot}G)^G = \mathcal{O}(B_{\cdot}G),$$

$$\mathcal{J}_\Delta(E_{\cdot}G)^G = \mathcal{J}_e(B_{\cdot}G).$$

Therefore, G being reductive,

$$\mathcal{O}(E_{\cdot}^{(1)}G)^G = \mathcal{O}(B_{\cdot}^{(1)}G). \tag{8.2}$$

REMARK 8.8. Indeed, it is not necessary to appeal to the reductivity of G to prove equation (8.2). The natural morphism

$$\mathcal{O}(E_nG)^G/\mathcal{J}_\Delta(E_nG)^G \to \left(\mathcal{O}(E_nG)/\mathcal{J}_\Delta(E_nG)\right)^G \tag{8.3}$$

is always injective. Thus, to prove equation (8.2), we only need to show the surjectivity of morphism (8.3). For each n, the principal G-bundle $\pi\colon E_nG \to B_nG$ is trivial. Therefore we can choose a section $s\colon B_nG \to E_nG$. But, if $f \in \mathcal{O}(E_nG)$ has its class $\overline{f} \in \left(\mathcal{O}(E_nG)/\mathcal{J}_\Delta(E_{\cdot}G)\right)^G$, then $\pi^*s^*(f) \in \mathcal{O}(E_nG)^G$ and $\overline{\pi^*s^*(f)} = \overline{f}$. Therefore the morphism (8.3) is surjective.

Let $\mathfrak{g}$ be the Lie algebra of G. By Theorem 8.4, there is a natural isomorphism

$$\overline{\psi}\colon \mathcal{N}\left(\mathcal{O}(E_{\cdot}^{(1)}G)\right) \to \Omega^{\cdot}(G),$$

where $\Omega^{\cdot}(G)$ is the space of algebraic differentials. By construction, this isomorphism is compatible with the right actions of G on $E_{\cdot}^{(1)}G$ and G. Therefore we obtain ([**2**, A 4.1]):

PROPOSITION 8.9. *There is a natural isomorphism*

$$\mathcal{N}\left(\mathcal{O}(B_{\cdot}^{(1)}G)\right) = \mathcal{N}\left(\mathcal{O}(E_{\cdot}^{(1)}G)\right)^G \to \Omega^{\cdot}(G)_R = E^{\cdot}(\mathfrak{g}, \mathbb{C}).$$

Recall that the Weil algebra has a bigrading given by

$$W^{p,q}(G) = S^p(\mathfrak{g}) \otimes E^{q-p}(\mathfrak{g}).$$

By construction, it is clear that $W^{*,\cdot}(G)$ satisfies the universal condition that defines $\Omega^*\left(E^{\cdot}(\mathfrak{g})\right)$. Thus both algebras are naturally isomorphic. Therefore, by Proposition 8.9, there is a natural isomorphism

$$\Omega^*\left(\mathcal{N}\left(\mathcal{O}(B_{\cdot}^{(1)}G)\right)\right) \to W^{*,\cdot}(G).$$

Hence, using Proposition 7.7 we obtain:

PROPOSITION 8.10. *There is a natural isomorphism*

$$\overline{\psi}\colon \mathcal{N}(\Omega^*(B^{(1)}_{\cdot}G)) = \mathcal{N}\Big(\overline{\Omega}^*\big(\mathcal{O}(B^{(1)}_{\cdot}G)\big)\Big) \to W^{*,\cdot}(G).$$

The isomorphism $\overline{\psi}$ induces a morphism

$$\psi\colon \mathcal{N}(\Omega^*(B.G)) \to W^{*,\cdot}(G).$$

Since the canonical connection is algebraic, the image $f_{\nabla_{E.G}}(I_G^*)$ lies in $\mathcal{N}(\Omega^*(B.G))$.

LEMMA 8.11. *The diagram*

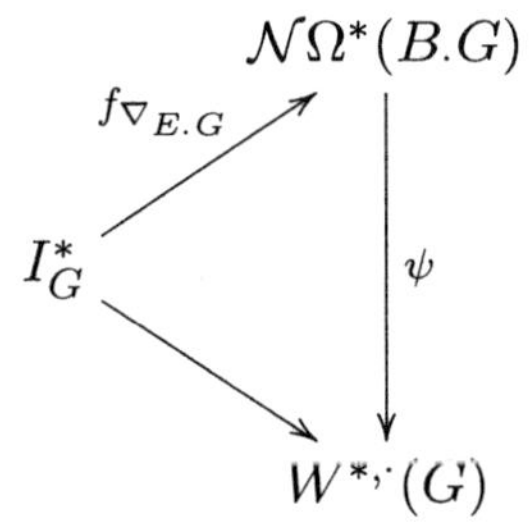

is commutative.

PROOF. By the splitting principle (see Remark 5.28) it is enough to check the case $G = \mathbb{G}_m$. This case follows easily from the definitions. □

Let us denote by F the filtration by the first degree of $\mathcal{N}(\Omega^*(B.G))$. That is

$$F^p\mathcal{N}(\Omega^*(B.G)) = \bigoplus_{p'\geq p} \mathcal{N}(\Omega^{p'}(B.G)).$$

Since $\psi\colon \mathcal{N}(\Omega^*(B.G)) \to W^{*,\cdot}(G)$ is a bigraded morphism, it is a filtered morphism with respect to the filtration F.

Since $B.G$ is a simplicial quasi-projective variety over $\mathbb{C}$, the cohomology of $B.G$ has a natural mixed Hodge structure. In particular there is a Hodge filtration F of $H^*(B.G, \mathbb{C})$. Moreover, it is known [**20**] that the mixed Hodge structure of $H^{2p}(B.G)$ is pure of type (p,p). Therefore

$$F^p\big(H^{2p}(B.G, \mathbb{C})\big) = H^{2p}(B.G, \mathbb{C}).$$

There is a natural map (that in general is not an isomorphism)

$$F^pH^{2p}(B.G, \mathbb{C}) \to H^{2p}\big(F^p\mathcal{N}\Omega^*(B.G)\big).$$

Composing this morphism with the morphism ψ, we obtain a morphism

$$H^{2p}(B.G, \mathbb{C}) \xrightarrow{\psi'} H^{2p}\big(F^pW(\mathfrak{g})\big) = I_G^{2p}.$$

From Lemma 8.11 it follows:

THEOREM 8.12. *The morphism ψ' is the inverse of the Chern–Weil morphism.*

8.3. A Description of the van Est Isomorphism

Let us begin this section by stating the analogue of Theorem 8.4 in the differentiable case. The proof is similar and will be omitted. Let X be a differentiable manifold. Then $E.X$ is a simplicial differentiable manifold and $E^0(E.X, \mathbb{R})$ is a cosimplicial algebra. We have that

$$E^0(E_nX, \mathbb{R}) = \overbrace{E^0(X, \mathbb{R}) \mathbin{\widehat{\otimes}_{\mathbb{R}}} \cdots \mathbin{\widehat{\otimes}_{\mathbb{R}}} E^0(X, \mathbb{R})}^{n+1}.$$

Let us write $I = I_\Delta = \operatorname{Ker}(\sigma^0) \subset E^0(E_1X, \mathbb{R})$ and let $J = J_\Delta$ be the cosimplicial ideal generated by I^2. The morphism

$$\begin{aligned} \psi_0^n \colon E^0(X, \mathbb{R}) \otimes \cdots \otimes E^0(X, \mathbb{R}) &\to E^n(X, \mathbb{R}) \\ f_0 \otimes f_1 \otimes \cdots \otimes f_n &\mapsto f_0\, df_1 \wedge \cdots \wedge df_n \end{aligned}$$

extends to a unique continuous linear morphism

$$\psi_0^n \colon E^0(E_nX, \mathbb{R}) \to E^n(X, \mathbb{R}).$$

THEOREM 8.13. *The sequence of morphisms ψ_0^n induces a morphism of differential graded algebras*

$$\psi \colon \mathcal{N}E^0(E.X, \mathbb{R}) \to E^{\cdot}(X, \mathbb{R})$$

and an isomorphism of differential graded algebras

$$\overline{\psi} \colon \mathcal{N}(E^0(E.X, \mathbb{R})/J) \to E^{\cdot}(X, \mathbb{R}).$$

Let now $X = G$ be a connected Lie group. The right action r of G on $E.G$ induces a left action of G on $\mathcal{N}E^0(E.G, \mathbb{R})$. The subcomplex of invariant elements is $\mathcal{N}E^0(E.G, \mathbb{R})^G = \mathcal{N}E^0(B.G, \mathbb{R})$. The ideal J is invariant under this action and $J_e = J^G$ is the cosimplicial ideal generated by $\operatorname{Ker}(\sigma^0)^2 \subset E^0(B_1G, \mathbb{R})$. Therefore, as in Remark 8.8,

$$\mathcal{N}(E^0(E.G, \mathbb{R})/J)^G = \mathcal{N}(E^0(B.G, \mathbb{R})/J_e).$$

We also consider the left action of G on $E^*(G, \mathbb{R})$ induced by the right action r. The subcomplex of invariant elements is the subalgebra of right invariant forms $E^*(G, \mathbb{R})_R$. With these actions the morphism ψ is a morphism of G-modules. Thus, if we restrict this morphism to the subcomplexes of invariant elements, we obtain:

PROPOSITION 8.14. *Let G be a connected Lie group, with Lie algebra $\mathfrak{g}$. Then the morphism ψ induces a natural isomorphism*

$$\mathcal{N}(E^0(B.G, \mathbb{R})/J_e) \to E^{\cdot}(\mathfrak{g}, \mathbb{R})$$

Now we can use the morphism ψ to give an explicit description of the van Est morphism ([**35**, III 7.3], [**2**, proof of Corollary A 5.2], [**55**, §3]).

THEOREM 8.15. *Let G be a connected Lie group, U a maximal compact subgroup of G and $\mathfrak{g}$ and $\mathfrak{u}$ the Lie algebras of G and U respectively. Then the induced morphism*

$$H^*_{\mathrm{cont}}(G,\mathbb{R}) = H^*(\mathcal{N}(E^0(E.G,\mathbb{R}))^G) \xrightarrow{H^*(\psi)} H^*(E^*(G,\mathbb{R})^G) = H^*(\mathfrak{g},\mathbb{R})$$

can be factored as

$$H^*_{\mathrm{cont}}(G,\mathbb{R}) \xrightarrow{\nu} H^*(\mathfrak{g},\mathfrak{u},\mathbb{R}) \xrightarrow{\iota} H^*(\mathfrak{g},\mathbb{R}),$$

where ν is the van Est isomorphism and ι is the obvious natural morphism.

PROOF. Let us write $M = U\backslash G$ and let $\pi\colon G \to M$ be the projection. There is a commutative diagram of G-modules

$$\begin{array}{ccc} \mathbb{R} & \longrightarrow & \mathcal{N}(E^0(E.G,\mathbb{R})) \\ \downarrow & & \downarrow \psi \\ E^{\cdot}(M,\mathbb{R}) & \xrightarrow{\pi^*} & E^{\cdot}(G,\mathbb{R}). \end{array}$$

The theorem follows from this commutative diagram taking continuous group cohomology. □

CHAPTER 9

Borel's Regulator

9.1. Algebraic K-Theory of Rings

In this section we will recall the definition of the algebraic K-theory of a ring. This section is based on [**58**]. In order to define Borel's regulator and to compare it with Beilinson's regulator for number fields, there is very little about K-theory that is needed. Only the relationship between K-theory and group homology that follows from the definition of K-theory as homotopy groups. Note however that, for the general construction of Beilinson's regulator map, and to establish its properties, it is useful to see algebraic K-theory as generalized sheaf cohomology ([**13**], cf. [**29**]). For more details about higher K-theory and Beilinson's regulator, the reader is referred to the appendixes of [**40**] and the bibliography thereof.

Let A be a commutative ring with unit. The groups $\mathrm{GL}_n(A)$ form a directed system as in Section 4.1. Let us write

$$\mathrm{GL}(A) = \varinjlim \mathrm{GL}_n(A).$$

The topological space $B\mathrm{GL}(A) = |\,B.\mathrm{GL}(A)|$ is a $K(\mathrm{GL}(A),1)$ space. The *plus construction* of $B\mathrm{GL}(A)$, denoted by $B\mathrm{GL}(A)^+$, is a topological space provided with a morphism

$$f\colon\ B\mathrm{GL}(A) \to B\mathrm{GL}(A)^+.$$

The space $B\mathrm{GL}(A)^+$ is characterized up to homotopy by the following properties:

(1) The map f is an acyclic cofibration. In particular $H_*(f)$ is an isomorphism.
(2) $\pi_1(B\mathrm{GL}(A)^+) = \mathrm{GL}(A)/[\mathrm{GL}(A), \mathrm{GL}(A)]$.

The space $B\mathrm{GL}(A)^+$ has a natural structure of H-space. In fact one can think of $B\mathrm{GL}(A)^+$ as a universal H-space associated to $B\mathrm{GL}(A)$.

Definition 9.1. Let A be a commutative ring with unity. The *algebraic K groups* of A are the homotopy groups of the topological space $B\mathrm{GL}(A)^+$:

$$K_m(A) = \pi_m(B\mathrm{GL}(A)^+).$$

Remark 9.2. Using the Hurewicz morphism and the fact that f is an acyclic morphism, we obtain a morphism

$$K_m(A) = \pi_m(BGL(A)^+, e)$$
$$\xrightarrow{\text{Hur}} H_m(BGL(A)^+, \mathbb{Z}) = H_m(BGL(A), \mathbb{Z}) = H_m(\mathrm{GL}(A), \mathbb{Z}),$$

where the last group is group homology.

Since $BGL(A)^+$ is an H-space, the results recalled in Chapter 3 imply that $H_*(BGL(A)^+, \mathbb{Z})$ has a structure of Hopf algebra. Moreover, if we denote by $P_m(BGL(A)^+, \mathbb{Q})$ the subspace of primitive elements, by Cartan–Serre Theorem (Theorem 3.17), the Hurewicz map induces an isomorphism

$$K_m(A) \otimes \mathbb{Q} \to P_m(BGL(A)^+, \mathbb{Q}) = P_m(\mathrm{GL}(A), \mathbb{Q}).$$

Therefore, a good knowledge of the homology of the group $\mathrm{GL}(A)$ will allow us to obtain information about the K-groups.

9.2. Definition of Borel's Regulator

For more details on the topics of this section and the next two sections, the reader is referred to the original paper of Borel [**8**] and the Bloch notes [**3**].

Let k be a number field and let $\mathfrak{o}$ be its ring of integers. For each $m \geq 2$, Borel's regulator map is a morphism

$$K_m(\mathfrak{o}) \to V_m,$$

from the K-theory of the ring $\mathfrak{o}$ to a certain real vector space. This map can be factored through the Hurewicz morphism

$$K_m(\mathfrak{o}) \xrightarrow{\text{Hur}} P_m(GL(\mathfrak{o}), \mathbb{Q}).$$

The group GL_n is reductive but not semisimple. In order to apply Theorem 9.6 below, Borel defined the regulator using the group SL_n instead of GL_n. The use of SL_n is justified (see [**4**, 3.3]) by the fact that the inclusion $\mathrm{SL}_n(\mathfrak{o}) \to \mathrm{GL}_n(\mathfrak{o})$ induces isomorphisms, for $m \geq 2$,

$$P_m(\mathrm{SL}_n(\mathfrak{o}), \mathbb{Q}) \to P_m(\mathrm{GL}_n(\mathfrak{o}), \mathbb{Q}).$$

Therefore the Hurewicz morphism also induces isomorphisms

$$K_m(\mathfrak{o}) \otimes \mathbb{Q} \to P_m(\mathrm{SL}(\mathfrak{o}), \mathbb{Q}), \quad \text{for } m \geq 2.$$

Let G be an algebraic group defined over $\mathbb{Q}$. Let us assume that $G(\mathbb{R})$ is a connected Lie group. Let Γ be a discrete group. Any group homomorphism $\varphi\colon \Gamma \to G(\mathbb{R})$ induces a morphism

$$\varphi^*\colon H^*_{\text{cont}}(G(\mathbb{R}), \mathbb{R}) \to H^*(\Gamma, \mathbb{R}).$$

Recall that in Section 6.3 we constructed an isomorphism

$$\gamma'\colon H^*\big(\mathrm{CT}\big(G(\mathbb{R})\big), \mathbb{R}\big) \to H^*_{\text{cont}}(G(\mathbb{R}), \mathbb{R}).$$

Thus, we obtain a morphism

$$j = \varphi^* \circ \gamma'\colon H^*\big(\mathrm{CT}\big(G(\mathbb{R})\big), \mathbb{R}\big) \to H^*(\Gamma, \mathbb{R}).$$

It is clear from the construction using differential forms that this morphism is a morphism of algebras.

Let us denote by $j^\vee$ the dual morphism

$$j^\vee : H_*(\Gamma, \mathbb{R}) \to H_*(\mathrm{CT}(G(\mathbb{R})), \mathbb{R}).$$

Since j is a morphism of algebras, $j^\vee$ is a morphism of coalgebras. Therefore it induces a morphism, also denoted $j^\vee$, between the primitive subspaces in homology

$$j^\vee : P_*(\Gamma, \mathbb{R}) \to P_*(\mathrm{CT}(G(\mathbb{R})), \mathbb{R}).$$

For each $n \geq 1$ let us write $G_n = \mathrm{Res}_{\mathbb{Q}/k} \mathrm{SL}_{n,k}$. That is, we view $\mathrm{SL}_{n,k}$ as an algebraic group over $\mathbb{Q}$. The subgroup $\mathrm{SL}_n(\mathfrak{v})$ is a discrete subgroup of $G_n(\mathbb{R})$. Let

$$j_n^\vee : H_*(\mathrm{SL}_n(\mathfrak{v}), \mathbb{R}) \to H_*(\mathrm{CT}(G_n(\mathbb{R})), \mathbb{R})$$

be as before. Let us denote

$$G(\mathbb{R}) = \lim_{n\to\infty} G_n(\mathbb{R}),$$
$$\mathrm{CT}(G(\mathbb{R})) = \lim_{n\to\infty} \mathrm{CT}(G_n(\mathbb{R})).$$

Taking the limit as n goes to infinity, we obtain a morphism

$$j^\vee : H_*(\mathrm{SL}(\mathfrak{v}), \mathbb{R}) \to H_*(\mathrm{CT}(G(\mathbb{R})), \mathbb{R}).$$

DEFINITION 9.3. Let $m \geq 2$. The mth *Borel regulator map* is the composition

$$r'_{\mathrm{Bo}} : K_m(\mathfrak{v}) \xrightarrow{\mathrm{Hur}} P_m(\mathrm{SL}(\mathfrak{v}), \mathbb{R}) \xrightarrow{j^\vee} P_m(\mathrm{CT}(G(\mathbb{R})), \mathbb{R}).$$

For the statements we will often need this process of taking the limit as n goes to infinity. But usually the proofs will be done for a fixed finite n. Since all the homology and cohomology theories that will appear satisfy a stability property, it is useful to have the following definition.

DEFINITION 9.4. Let n and m be positive integers. We will say that n, m are in *the stable range* if n is odd and $m \leq (n-1)/2$.

Observe that the condition $m \leq (n-1)/2$ ensures that the morphisms

$$H_m(\mathrm{SL}_n(\mathfrak{v}), \mathbb{Q}) \to H_m(\mathrm{SL}(\mathfrak{v}), \mathbb{Q}),$$
$$H^m(\mathrm{SL}(\mathbb{C}), \mathbb{R}) \to H^m(\mathrm{SL}_n(\mathbb{C}), \mathbb{R}),$$

and the corresponding morphism for $\mathrm{SL}_n(\mathbb{R})$ are isomorphisms. The condition n odd implies that the inclusion $\mathfrak{so}_n(\mathbb{R}) \to \mathfrak{u}_n$ is noncohomologous to zero. Therefore the morphism

$$H^*(\mathfrak{u}_n, \mathfrak{so}_n(\mathbb{R})) \to H^*(\mathfrak{u}_n)$$

is injective.

In order to know the behavior of Borel's regulator map, we need more information about the morphism $j^\vee$. This is provided by Borel's theory of arithmetic groups. Let G be a closed subgroup of $\mathrm{GL}(V)$, for V a finite dimensional $\mathbb{Q}$-vector space. Let L be a lattice of V. Let us write G_L for the subgroup of $G(\mathbb{Q})$ that leaves L fixed:

$$G_L = \{g \in G(\mathbb{Q}) \mid g(L) = L\}.$$

DEFINITION 9.5. A subgroup Γ of $G(\mathbb{Q})$ is called arithmetic if it is commensurable with G_L. That is, if $\Gamma \cap G_L$ has finite index in Γ and in G_L.

For the properties of arithmetic subgroups the reader is referred to [**6**]. For our purposes the main result of the theory is the following.

THEOREM 9.6 (Borel [**7**]). *Let G be a semisimple algebraic group defined over $\mathbb{Q}$ and let Γ be an arithmetic subgroup of $G(\mathbb{R})$, with $\varphi\colon \Gamma \to G(\mathbb{R})$ the inclusion. Let $\mathrm{CT}\big(G(\mathbb{R})\big)$ be the compact twin of $G(\mathbb{R})$ and j as before. Then there exists a number $\rho(G)$ such that*

$$j\colon H^m\big(\mathrm{CT}\big(G(\mathbb{R})\big), \mathbb{R}\big) \to H^m(\Gamma, \mathbb{R})$$

is an isomorphism for $m \leq \rho(G)$. This number $\rho(G)$ is computable from the algebraic structure of G.

REMARK 9.7. For any given Lie group G, the number $\rho(G)$ may be small. For instance $\rho(\mathrm{SL}_n) = n/4$. But, as the rank of G goes to infinity, the number $\rho(\mathrm{SL}_n)$ also goes to infinity. Thus this result allows us to compute the stable real cohomology groups of the arithmetic groups.

COROLLARY 9.8. *Let $m \geq 2$ be an integer. Then the Borel regulator map induces an isomorphism*

$$K_m(\mathfrak{v}) \otimes \mathbb{R} \to P_m\big(\mathrm{CT}\big(G(\mathbb{R})\big), \mathbb{R}\big).$$

9.3. The Rank of the Groups $K_m(\mathfrak{v})$

Let k and $\mathfrak{v}$ be as in the last section. Let Σ be the set of complex immersions of k, and let $\mathcal{V}$ be the set of Archimedean places of k. We will use $\mathcal{V}_\mathbb{R}$ to refer to the set of places corresponding to the real immersions and $\mathcal{V}_\mathbb{C}$ to refer to the set of places corresponding to the complex immersions. Let $d = [k : \mathbb{Q}]$ be the degree of k. As usual, we write $r_1 = \sharp\mathcal{V}_\mathbb{R}$ and $r_2 = \sharp\mathcal{V}_\mathbb{C}$. Then $d = r_1 + 2r_2$. For a fixed m let us choose an integer n such that n, m are in the stable range.

As in the previous section, let us write $G_n = \mathrm{Res}_{\mathbb{Q}/k} \mathrm{SL}_{n,k}$. Then the Lie group $G_n(\mathbb{R})$ is

$$G_n(\mathbb{R}) = \prod_{\nu \in \mathcal{V}_\mathbb{R}} \mathrm{SL}_n(\mathbb{R}) \times \prod_{\nu \in \mathcal{V}_\mathbb{C}} \mathrm{SL}_n(\mathbb{C}).$$

In order to construct the compact twin of this group we can work componentwise. For the group $\mathrm{SL}_n(\mathbb{R})$ we have that the maximal compact subgroup is $\mathrm{SO}_n(\mathbb{R})$. The complexification is $\mathrm{SL}_n(\mathbb{C})$. The compact subgroup of the complexification is SU_n. Therefore the compact twin is $\mathrm{CT}(\mathrm{SL}_n(\mathbb{R})) = \mathrm{SO}_n(\mathbb{R})\backslash \mathrm{SU}_n$.

For the group $\mathrm{SL}_n(\mathbb{C})$ the maximal compact subgroup is SU_n. The complexification is $\mathrm{SL}_n(\mathbb{C}) \times \mathrm{SL}_n(\mathbb{C})$, with the inclusion

$$\begin{aligned} \mathrm{SL}_n(\mathbb{C}) &\to \mathrm{SL}_n(\mathbb{C}) \times \mathrm{SL}_n(\mathbb{C}) \\ M &\mapsto (\overline{M}, M) \end{aligned}$$

The complex conjugation $\tau\colon \mathrm{SL}_n(\mathbb{C}) \times \mathrm{SL}_n(\mathbb{C}) \to \mathrm{SL}_n(\mathbb{C}) \times \mathrm{SL}_n(\mathbb{C})$ is given by $\tau(M, N) = (\overline{N}, \overline{M})$. Thus the compact subgroup of the complexification is $\mathrm{SU}_n \times \mathrm{SU}_n$ and the compact twin is homeomorphic to SU_n.

Therefore, after taking the limit when n goes to infinity, the compact twin $\mathrm{CT}\big(G(\mathbb{R})\big)$ is homeomorphic to $(\mathrm{SO} \setminus \mathrm{SU})^{r_1} \times \mathrm{SU}^{r_2}$. Hence, Theorem 9.6, Corollary 4.27 and Corollary 4.23 imply

THEOREM 9.9 (Borel). *The rank of the groups $K_m(\mathfrak{v})$, $m \geq 2$, are*

$$\mathrm{rk}\big(K_m(\mathfrak{v})\big) = \begin{cases} 0, & \text{if } m \text{ is even,} \\ r_1 + r_2, & \text{if } m \equiv 1 \bmod 4, \\ r_2, & \text{if } m \equiv 3 \bmod 4. \end{cases}$$

9.4. The Values of the Zeta Functions

In this section we will recall Borel's Theorem relating the regulator with the values of zeta functions.

The main reason for using $P_{2p-1}\big(\mathrm{CT}\big(G(\mathbb{R})\big), \mathbb{R}\big)$ as the target for the regulator map is that it has a natural integral structure. Since K-theory is defined using homotopy, Lichtenbaum [**42**] proposed to use the lattice given by the image of $\pi_{2p-1}(\mathrm{CT}\big(G(\mathbb{R})\big), e)$ under the Hurewicz morphism. Let us denote by L'_{2p-1} this lattice. This is also the lattice used in [**8**, §6.4] to define the regulator number.

DEFINITION 9.10. The *Borel regulator*, $R'_{\mathrm{Bo},p}$, is the covolume of the lattice

$$r'_{\mathrm{Bo}}\big(K_{2p-1}(\mathfrak{v})\big) \subset P_{2p-1}\big(\mathrm{CT}\big(G(\mathbb{R})\big), \mathbb{R}\big)$$

with respect to the lattice L'_{2p-1}.

Let us write

$$d_p = \mathrm{rk}\big(K_{2p-1}(\mathfrak{v})\big) = \dim\Big(P_{2p-1}\big(\mathrm{CT}\big(G(\mathbb{R})\big), \mathbb{R}\big)\Big).$$

Let $\zeta_k(s)$ be the Dedekind zeta function of the number field k.

Lichtenbaum ([**42**]) proposed the following question, which is known as Lichtenbaum's conjecture.

QUESTION 9.11. When is it true that

$$R'_{\mathrm{Bo},p} = \pm \frac{\sharp K_{2p-1}(\mathfrak{v})_{\mathrm{torsion}}}{\sharp K_{2p-2}(\mathfrak{v})} \lim_{s \to -p+1} \zeta_k(s)(s+p-1)^{-d_p}?$$

When asking this question, Lichtenbaum remarked that, due to the lack of known examples at that time, the normalization may not be the correct one and that the formula might need to be adjusted by some power of π and some rational number.

Later Borel ([**8**], see also [**3**]) proved the transcendental part of Lichtenbaum question. Given two real numbers a and b we will write $a \sim b$ if there is a non zero rational number q with $a = qb$.

THEOREM 9.12 (Borel). *Let k be a number field. Then*

$$R'_{\mathrm{Bo},p} \sim \pi^{-d_p} \lim_{s \to -p+1} \zeta_k(s)(s+p-1)^{-d_p}. \tag{9.1}$$

Observe that the factor π^{-d_p} implies that the lattice chosen is not the best one. This will be one of the reasons for renormalizing Borel's regulator.

REMARK 9.13. The value of Borel's regulator is determined by the integral structure given by the lattice L'_{2p-1}. Therefore, for any real vector space V, any lattice L of V and any isomorphism $f\colon P_{2p-1}(\mathrm{CT}(G(\mathbb{R})), \mathbb{R}) \to V$ with $f(L'_{2p-1}) = L$, we will say that the composition $f \circ r'_{\mathrm{Bo}}$ is equivalent to the Borel regulator map.

9.5. A Renormalization of Borel's Regulator

The definition of Borel's regulator given in Section 9.2 is the original definition which appears in the paper [**8**]. Nevertheless, in the literature there appears a slightly different morphism with the name of Borel's regulator ([**2, 23, 25, 34, 55**]). The aim of this section is fill the gap between the two definitions. The first difference is obvious and the reason for it is to avoid the factor π^{-d_p} in equation (9.1). The second difference is related with the choice of the lattice L' and the reason for it is the different behavior of the homotopy groups of the compact twins of $\mathrm{SL}(\mathbb{R})$ and $\mathrm{SL}(\mathbb{C})$. As we will see, this behavior implies that Borel's regulator, as defined, does not factorize in a uniform way through the K-theory of the field $\mathbb{C}$.

A key ingredient in the definition of Borel's regulator is the isomorphism

$$\gamma'\colon H^*(\mathrm{CT}(G_n(\mathbb{R})), \mathbb{R}) \to H^*_{\mathrm{cont}}(G_n(\mathbb{R}), \mathbb{R}).$$

We have used this isomorphism only for odd degrees. Let us denote by $\mathfrak{g}_{\mathbb{R}}$, $\mathfrak{k}$ and $\mathfrak{g}_u$ the Lie algebras of the group $G_n(\mathbb{R})$, of the maximal compact subgroup $K \subset G_n(\mathbb{R})$ and of the compact subgroup $G_{n,u} = \big(G_n(\mathbb{R})\big)_u \subset G_n(\mathbb{C})$ respectively. Recall that the isomorphism γ' is the composition

$$\begin{aligned} H^{2p-1}(\mathrm{CT}(G_n(\mathbb{R})), \mathbb{R}) &\xrightarrow{\alpha} H^{2p-1}(\mathfrak{g}_u, \mathfrak{k}, \mathbb{R}) \\ &\xrightarrow{\iota} H^{2p-1}(\mathfrak{g}_{\mathbb{R}}, \mathfrak{k}, \mathbb{R}) \\ &\xrightarrow{\beta} H^{2p-1}_{\mathrm{cont}}(G_n(\mathbb{R}), \mathbb{R}), \end{aligned}$$

where β is the inverse of the van Est isomorphism, the isomorphism α is constructed by means of invariant differential forms and ι is the isomorphism given by multiplication with i^{2p-1}. The first change we will make to the regulator is to keep a twist and replace the isomorphism ι by the identity

$$E^{2p-1}\big(\mathfrak{g}_{\mathbb{R}}, \mathfrak{k}, \mathbb{R}(p-1)\big) = E^{2p-1}\big(\mathfrak{g}_u, \mathfrak{k}, \mathbb{R}(p)\big),$$

where $\mathbb{R}(p)$ is the subgroup $(2\pi i)^p\mathbb{R} \subset \mathbb{C}$. Let γ be the composition

$$\begin{aligned} H^{2p-1}\big(\mathrm{CT}\big(G_n(\mathbb{R})\big), \mathbb{R}(p)\big) &\xrightarrow{\alpha} H^{2p-1}\big(\mathfrak{g}_u, \mathfrak{k}, \mathbb{R}(p)\big) \\ &= H^{2p-1}\big(\mathfrak{g}_{\mathbb{R}}, \mathfrak{k}, \mathbb{R}(p-1)\big) \\ &\xrightarrow{\beta} H^{2p-1}_{\mathrm{cont}}\big(G_n(\mathbb{R}), \mathbb{R}(p-1)\big), \end{aligned}$$

Twisting the source and the target of γ by $\mathbb{R}(1-p)$ and composing with φ^* as in Section 9.2 we obtain a morphism

$$j''\colon H^{2p-1}\big(\mathrm{CT}\big(G_n(\mathbb{R})\big), \mathbb{R}(1)\big) \to H^{2p-1}(\mathrm{SL}_n(\mathfrak{v}), \mathbb{R}).$$

Taking duals and restricting to the primitive subspaces we get a morphism

$$(j'')^\vee\colon P_{2p-1}(\mathrm{SL}_n(\mathfrak{v}), \mathbb{R}) \to P_{2p-1}\big(\mathrm{CT}\big(G_n(\mathbb{R})\big), \mathbb{R}(-1)\big).$$

Taking the limit as n goes to infinity and composing with the Hurewicz morphism we obtain a morphism

$$r''_{\mathrm{Bo}}\colon K_{2p-1}(\mathfrak{v}) \to P_{2p-1}\big(\mathrm{CT}\big(G(\mathbb{R})\big), \mathbb{R}(-1)\big).$$

In $P_{2p-1}\big(\mathrm{CT}\big(G(\mathbb{R})\big), \mathbb{R}(-1)\big)$ we put the lattice

$$L''_{2p-1} = \pi_{2p-1}\big(\mathrm{CT}\big(G(\mathbb{R})\big), e\big)(-1) = (2\pi i)^{-1}\pi_{2p-1}\big(\mathrm{CT}\big(G(\mathbb{R})\big), e\big).$$

REMARK 9.14. With this lattice, the morphism r''_{Bo} is not equivalent to r'_{Bo} but to $2\pi r'_{\mathrm{Bo}}$. Thus this change will take care of the factor π^{-d_p} in Theorem 9.12.

In order to try to factor r''_{Bo} through the K-theory of $\mathbb{C}$, let us consider $G_n(\mathbb{C})$ as a real Lie group. Its compact twin is homeomorphic to SU_n^Σ. As before, the morphisms $\mathrm{SL}_n(\mathfrak{v}) \to G_n(\mathbb{C})$, for all n, induce morphisms

$$\rho\colon K_{2p-1}(\mathfrak{v}) \to P_{2p-1}\big(\mathrm{SU}^\Sigma, \mathbb{R}(-1)\big).$$

Let F be the involution of $P_{2p-1}\big(\mathrm{SU}^\Sigma, \mathbb{R}(-1)\big)$ that acts as complex conjugation on the space SU^Σ and on the coefficients at the same time. Note that complex conjugation on the space SU^Σ means complex conjugation in each factor SU, and complex conjugation in the set of complex immersions Σ. Let us denote by $P_{2p-1}\big(\mathrm{SU}^\Sigma, \mathbb{R}(-1)\big)^F$ the subspace of invariant elements. We will use the same letter F in any circumstance when such an involution can be constructed.

The morphism of real Lie groups $G_n(\mathbb{R}) \to G_n(\mathbb{C})$ induces a morphism of compact twins

$$\psi\colon \mathrm{CT}\big(G(\mathbb{R})\big) = (\mathrm{SO}\setminus\mathrm{SU})^{r_1} \times \mathrm{SU}^{r_2} \to \mathrm{CT}\big(G(\mathbb{C})\big) = \mathrm{SU}^\Sigma.$$

PROPOSITION 9.15. *The morphism*

$$\psi_*\colon H_{2p-1}\big(\mathrm{CT}\big(G(\mathbb{R})\big),\mathbb{R}(-1)\big) \to H_{2p-1}\big(\mathrm{CT}\big(G(\mathbb{C})\big),\mathbb{R}(-1)\big)$$

induces an isomorphism

$$\psi_*\colon P_{2p-1}\big(\mathrm{CT}\big(G(\mathbb{R})\big),\mathbb{R}(-1)\big) \xrightarrow{\cong} P_{2p-1}\big(\mathrm{CT}\big(G(\mathbb{C})\big),\mathbb{R}(-1)\big)^F.$$

PROOF. It is enough to treat the cases $r_1 = 1$, $r_2 = 0$ and $r_1 = 0$, $r_2 = 1$.

Let us prove the first case. If $r_1 = 1$ and $r_2 = 0$ then $G(\mathbb{R}) = \mathrm{SL}(\mathbb{R})$ and $G(\mathbb{C}) = \mathrm{SL}(\mathbb{C})$. The inclusion $\mathrm{SL}(\mathbb{R}) \to \mathrm{SL}(\mathbb{C})$ induces a commutative diagram of compact subgroups and compact twins

$$\begin{array}{ccccc} \mathrm{SO} & \longrightarrow & \mathrm{SU} & \longrightarrow & \mathrm{SO}\setminus\mathrm{SU} \\ \downarrow & & \downarrow{\scriptstyle\alpha} & & \downarrow{\scriptstyle\psi} \\ \mathrm{SU} & \xrightarrow{\beta} & \mathrm{SU}\times\mathrm{SU} & \xrightarrow{\varphi} & \mathrm{SU}, \end{array} \tag{9.2}$$

where $\alpha(M) = (M, M)$, whereas $\beta(M) = (\overline{M}, M)$. Thus

$$\varphi(M, N) = \overline{M}^{-1}N = M^tN$$

and $\psi(M) = M^tM$. Since the rows are fibrations, the homotopy long exact sequence of a fibration gives rise to a morphism of exact sequences

$$\begin{array}{ccccc} P_*\big(\mathrm{SO},\mathbb{R}(-1)\big) & \longrightarrow & P_*\big(\mathrm{SU},\mathbb{R}(-1)\big) & \longrightarrow & P_*\big(\mathrm{SO}\setminus SU,\mathbb{R}(-1)\big) \\ \downarrow & & \downarrow & & \downarrow \\ P_*\big(\mathrm{SU},\mathbb{R}(-1)\big) & \longrightarrow & P_*\big(\mathrm{SU}\times SU,\mathbb{R}(-1)\big) & \longrightarrow & P_*\big(\mathrm{SU},\mathbb{R}(-1)\big), \end{array}$$

By Theorem 4.26 we know that $P_m\big(\mathrm{SO},\mathbb{R}(-1)\big) = \mathbb{R}(-1)$ if $m \equiv 3 \bmod 4$ and zero otherwise, and that $P_m\big(\mathrm{SO}\setminus\mathrm{SU},\mathbb{R}(-1)\big) = \mathbb{R}(-1)$ if $m \equiv 1 \bmod 4$ and zero otherwise.

LEMMA 9.16. *Let $s\colon \mathrm{SU} \to \mathrm{SU}$ be the morphism given by complex conjugation. Then the induced morphism*

$$s_{2p-1}\colon P_{2p-1}\big(\mathrm{SU},\mathbb{R}(-1)\big) \to P_{2p-1}\big(\mathrm{SU},\mathbb{R}(-1)\big)$$

is equal to $(-1)^p$.

PROOF. This follows from the explicit description of the generators of the group $P_{2p-1}\big(\mathrm{SU},\mathbb{R}(-1)\big)$ given in Definition 4.5 and the fact that complex conjugation reverses the orientation of $\mathbb{C}^p$ if p is odd and preserves the orientation if p is even. □

Observe that in this lemma the morphism s acts as complex conjugation on the space and not on the coefficients. Thus it does not agree with F, in fact, since the coefficients are $\mathbb{R}(-1)$, $F = -s$. Therefore, if p is even $P_{2p-1}\big(\mathrm{SU},\mathbb{R}(-1)\big)^F = 0$, proving the proposition in this case.

Let p be an odd number. We can identify $P_{2p-1}(\mathrm{SU},\mathbb{R}) \oplus P_{2p-1}(\mathrm{SU},\mathbb{R})$ with $P_{2p-1}(\mathrm{SU}\times\mathrm{SU},\mathbb{R})$, by the isomorphism defined by the two projections of $\mathrm{SU}\times\mathrm{SU}$. By Lemma 9.16 we have that, for $x\in P_{2p-1}(\mathrm{SU},\mathbb{R})$, $\beta_*(x)=(-x,x)$, whereas $\alpha_*(x)=(x,x)$. Therefore, if $x\neq 0$, $\varphi_*\big(\alpha_*(x)\big)\neq 0$. Hence ψ_* is injective. But by Lemma 9.16, $P_m\big(\mathrm{SU},\mathbb{R}(-1)\big)^F = P_m\big(\mathrm{SU},\mathbb{R}(-1)\big)$ has dimension one. This proves the case p odd.

If $r_1=0$ and $r_2=1$, we have that $G(\mathbb{R})=\mathrm{SL}(\mathbb{C})$ and $G(\mathbb{C})=\mathrm{SL}(\mathbb{C})\times\mathrm{SL}(\mathbb{C})$. Thus in this case the analogue of diagram (9.2) is

$$\begin{array}{ccccc} \mathrm{SU} & \longrightarrow & \mathrm{SU}\times\mathrm{SU} & \longrightarrow & \mathrm{SU} \\ \downarrow & & \downarrow{\scriptstyle\alpha} & & \downarrow{\scriptstyle\psi} \\ \mathrm{SU}\times\mathrm{SU} & \xrightarrow{\beta} & \mathrm{SU}\times\mathrm{SU}\times\mathrm{SU}\times\mathrm{SU} & \xrightarrow{\varphi} & \mathrm{SU}\times\mathrm{SU}, \end{array} \tag{9.3}$$

with

$$\begin{aligned} \alpha(M,N) &= (M,N,M,N),\\ \beta(M,N) &= (\overline{N},\overline{M},M,N),\\ \varphi(A,B,M,N) &= (B^tM,A^tN) \end{aligned}$$

and

$$\psi(M)=(M^t,M).$$

The result follows as in the previous case. □

COROLLARY 9.17. *The image $\rho\big(K_{2p-1}(\mathfrak{v})\big)$ lies in the subgroup of F-invariant elements $P_{2p-1}\big(\mathrm{CT}\big(G(\mathbb{C})\big),\mathbb{R}(-1)\big)^F$. Moreover there is a commutative diagram with the vertical arrow being an isomorphism,*

$$\begin{array}{ccc} & & P_{2p-1}\big(\mathrm{CT}\big(G(\mathbb{R})\big),\mathbb{R}(-1)\big) \\ & \overset{r''_{\mathrm{Bo}}}{\nearrow} & \\ K_{2p-1}(\mathfrak{v}) & & \downarrow{\scriptstyle\psi_*} \\ & \underset{\rho}{\searrow} & \\ & & P_{2p-1}\big(\mathrm{CT}\big(G(\mathbb{C})\big),\mathbb{R}(-1)\big)^F \end{array}$$

This corollary means that, for computing the rank, it is the same to use ρ or r''_{Bo}.

Let us see what happens to the integral structure. The natural lattice to consider in the real vector space $P_{2p-1}\big(\mathrm{SU}^\Sigma,\mathbb{R}(-1)\big)^F$ is

$$L_{2p-1}=\big(\pi_{2p-1}(\mathrm{SU}^\Sigma,e)(-1)\big)^F.$$

In order to compare the lattices L''_{2p-1} and L_{2p-1}, we have to study the effect in homotopy of the morphism $\mathrm{CT}\big(G(\mathbb{R})\big)\to\mathrm{CT}\big(G(\mathbb{C})\big)$. It is enough to treat the cases $r_1=1$, $r_2=0$ and $r_1=0$, $r_2=1$.

LEMMA 9.18. (1) *The morphism* $\mathrm{CT}(\mathrm{SL}(\mathbb{R})) \to \mathrm{CT}(\mathrm{SL}(\mathbb{C}))$ *sends a generator of the group* $\pi_{8k+1}(\mathrm{SO}\setminus\mathrm{SU}, e)$ *to a generator of* $\pi_{8k+1}(\mathrm{SU}, e)$ *and sends a generator of* $\pi_{8k+5}(\mathrm{SO}\setminus\mathrm{SU}, e)$ *to twice a generator of* $\pi_{8k+5}(\mathrm{SU}, e)$.

(2) *For each* p *let us denote by* ϵ_{2p-1} *a generator of* $\pi_{2p-1}(SU, e)$. *Then the morphism* $\mathrm{CT}(\mathrm{SL}(\mathbb{C})) \to \mathrm{CT}(\mathrm{SL}(\mathbb{C})\times\mathrm{SL}(\mathbb{C}))$ *sends* ϵ_{2p-1} *to the element* $((-1)^{p+1}\epsilon_{2p-1}, \epsilon_{2p-1})$.

PROOF. Using Bott's Periodicity Theorem (Theorem 4.26), and the homotopy long exact sequence of the fibration

$$\mathrm{SO} \to \mathrm{SU} \to \mathrm{SO}\setminus\mathrm{SU},$$

one can see that the morphism $\pi_m(\mathrm{SU}, e) \to \pi_m(\mathrm{SU}/\mathrm{SO}, e)$ is surjective if $m \equiv 5 \bmod 8$ and has cokernel $\mathbb{Z}/2\mathbb{Z}$ if $m \equiv 1 \bmod 8$. Thus case (1) follows from this and from the morphism of homotopy long exact sequences associated to diagram (9.2).

Case (2) follows from the morphism of homotopy long exact sequences associated to diagram (9.3). □

DEFINITION 9.19. The (*renormalized*) *Borel regulator* map is the map

$$K_{2p-1}(\mathfrak{v}) \xrightarrow{\rho} P_{2p-1}\big(\mathrm{CT}(G(\mathbb{C})), \mathbb{R}(-1)\big)^F,$$

where the space in the right hand side is provided with the lattice

$$L_{2p-1} = \big(\pi_{2p-1}(\mathrm{CT}(G(\mathbb{C})), e)(-1)\big)^F.$$

This map will be denoted by r_{Bo}. We denote by $R_{\mathrm{Bo},p}$ the covolume of the lattice $r_{\mathrm{Bo}}(K_{2p-1}(\mathfrak{v}))$ with respect to the lattice L_{2p-1}.

REMARK 9.20. If the field k does not have real immersions, Lemma 9.18 implies that r_{Bo} and r''_{Bo} are equivalent. Therefore the factor 2π is the only discrepancy between the two definitions of Borel's regulator. On the other hand, if the field has real immersions, by Lemma 9.18 there is also a difference by a factor of 2 in the component corresponding with the real immersions, when $p \equiv 3 \bmod 4$. Therefore, we have

$$R_{\mathrm{Bo},p} = (2\pi)^{d_p} R'_{\mathrm{Bo},p}, \quad \text{if } p \not\equiv 3 \bmod 4, \tag{9.4}$$

$$R_{\mathrm{Bo},p} = (2\pi)^{d_p} 2^{r_1} R'_{\mathrm{Bo},p} \quad \text{if } p \equiv 3 \bmod 4. \tag{9.5}$$

Let us see that the renormalized Borel regulator map can be factored through $K(\mathbb{C})$. The group $\mathrm{SL}_n(\mathbb{C})$ with the discrete topology will be denoted by $\mathrm{SL}_n(\mathbb{C})^\delta$. Applying the process that defines the renormalized Borel regulator map to the natural morphism

$$\mathrm{SL}_n(\mathbb{C})^\delta \to \mathrm{SL}_n(\mathbb{C})$$

induces, for $p \geq 2$, a map

$$K_{2p-1}(\mathbb{C}) \to P_{2p-1}\big(\mathrm{SU}, \mathbb{R}(-1)\big)$$

denoted by $r_{\mathrm{Bo},\mathbb{C}}$. The following result is clear.

PROPOSITION 9.21. *Let k be a number field, $\mathfrak{v}$ its ring of integers and Σ the set of complex immersions of k. For each $\sigma \in \Sigma$, let us also denote by σ the induced morphism $\sigma\colon K_{2p-1}(\mathfrak{v}) \to K_{2p-1}(\mathbb{C})$. Then the renormalized Borel regulator map can be factored as*

$$K_{2p-1}(\mathfrak{v}) \xrightarrow{\prod_\Sigma \sigma} \Big(\prod_\Sigma K_{2p-1}(\mathbb{C})\Big)^F \xrightarrow{\prod r_{\mathrm{Bo},\mathbb{C}}} \Big(\prod_\Sigma P_{2p-1}(SU,\mathbb{R}(-1))\Big)^F.$$

The last objective of this section is to give an equivalent definition of the renormalized Borel regulator that hides the fact that the integral lattice we are considering is given by the homotopy of the compact twin. The key ingredient for this alternative definition is the Chern character.

A class $\alpha \in H^{2p-1}(\mathrm{SU},\mathbb{R}(p))$ determines a morphism

$$\alpha\colon H_{2p-1}(\mathrm{SU},\mathbb{R}(-1)) \to \mathbb{R}(p-1).$$

Hence, by restriction, it determines a morphism

$$\alpha\colon P_{2p-1}(\mathrm{SU},\mathbb{R}(-1)) \to \mathbb{R}(p-1).$$

Moreover, if the class α is invariant under F, then we obtain a morphism

$$\alpha\colon P_{2p-1}(\mathrm{SU}^\Sigma,\mathbb{R}(-1))^F \to \Big(\prod_{\sigma\in\Sigma} \mathbb{R}(p-1)\Big)^F.$$

Let ch_p be the component in $H^{2p}(B\mathrm{SU},\mathbb{R}(p))$ of the twisted Chern character of the universal bundle. Let $\mathfrak{s}$ denote the suspension. By Remark 4.25 we know that, if ε_{2p-1} is a generator of $\pi_{2p-1}(SU,e)$ and Hur denotes the Hurewicz morphism then $\mathfrak{s}(\mathrm{ch}_p)(\mathrm{Hur}(\varepsilon_{2p-1}))$ is a generator of $\mathbb{Z}(p) \subset \mathbb{R}(p)$. Moreover, $\mathfrak{s}(\mathrm{ch}_p)$ is invariant under F. Thus if we add to the real vector space $\left(\prod \mathbb{R}(p-1)\right)^F$ the integral structure given by $\left(\prod \mathbb{Z}(p-1)\right)^F$, we have:

PROPOSITION 9.22. *The morphism*

$$\prod_\Sigma \mathfrak{s}(\mathrm{ch}_p) : \pi_{2p-1}(SU^\Sigma,e)\otimes\mathbb{R}(-1) \to \prod_\Sigma \mathbb{R}(p-1)$$

is an isomorphism. Moreover it is compatible with the integral structure and with the involution F.

COROLLARY 9.23. *The composition*

$$\begin{aligned} K_{2p-1}(\mathfrak{v}) &\xrightarrow{\prod_\Sigma \sigma} \Big(\prod_\Sigma K_{2p-1}(\mathbb{C})\Big)^F \\ &\xrightarrow{\prod_\Sigma r_{\mathrm{Bo},\mathbb{C}}} \Big(\prod_\Sigma P_{2p-1}(\mathrm{SU},\mathbb{R}(-1))\Big)^F \\ &\xrightarrow{\prod_\Sigma \mathfrak{s}(\mathrm{ch}_p)} \Big(\prod_\Sigma \mathbb{R}(p-1)\Big)^F. \end{aligned}$$

is equivalent to the renormalized Borel regulator map.

9.6. Borel Elements

Recall that we have defined algebraic K-theory using the general linear group GL. But in the definition of the Borel regulator we shifted to the special linear group SL in order to apply Theorem 9.6. Since the inclusion $\mathrm{SL}_n \to \mathrm{GL}_n$ induces isomorphisms, for $m \geq 2$

$$P_m(\mathrm{SL}_n(\mathfrak{v}), \mathbb{R}) \cong P_m(\mathrm{GL}_n(\mathfrak{v}), \mathbb{R}),$$
$$\pi_m(\mathrm{SU}_n, e) \cong \pi_m(\mathrm{U}_n, e),$$

we can replace, in the definition of the renormalized Borel regulator, the group SL_n with the group GL_n.

By Proposition 9.21 Borel's regulator for number fields is determined by Borel's regulator for the complex field. By Corollary 9.23 Borel's regulator for the complex field is equivalent to the composition

$$K_{2p-1}(\mathbb{C}) \xrightarrow{\mathrm{Hur}} P^{\mathrm{group}}_{2p-1}(\mathrm{GL}(\mathbb{C}), \mathbb{R}) \to \mathbb{R}(p-1),$$

where $P^{\mathrm{group}}_{2p-1}$ denote the primitive part of group homology. Therefore, Borel's regulator is determined by a map between the primitive part of the homology groups $P^{\mathrm{group}}_{2p-1}(GL(\mathbb{C}), \mathbb{R})$ and the real vector space $\mathbb{R}(p-1)$. For the sake of comparison, it is useful to work, not with the map from homology, but with the corresponding element in cohomology. By construction, we know that this element comes from continuous cohomology. The Borel element will be this element in continuous cohomology. Let us choose n such that $n, 2p$ are in the stable range.

DEFINITION 9.24. Let $\mathrm{ch}_p \in H^{2p}\big(B.\mathrm{GL}_n(\mathbb{C}), \mathbb{R}(p)\big)$ be the p-th component of the twisted Chern character. Then the *Borel element*, Bo_p is the image of ch_p under the composition

$$\begin{aligned} H^{2p}\big(B.\mathrm{GL}_n(\mathbb{C}), \mathbb{R}(p)\big) &\xrightarrow{\mathfrak{s}} H^{2p-1}\big(\mathrm{GL}_n(\mathbb{C}), \mathbb{R}(p)\big) \\ &= H^{2p-1}\big(\mathrm{U}_n, \mathbb{R}(p)\big) \\ &\xrightarrow{\alpha} H^{2p-1}\big(\mathfrak{u}_n \oplus \mathfrak{u}_n, \mathfrak{u}_n, \mathbb{R}(p)\big) \\ &\xrightarrow{\mathrm{Id}} H^{2p-1}\big(\mathfrak{gl}_n(\mathbb{C}), \mathfrak{u}_n, \mathbb{R}(p-1)\big) \\ &\xrightarrow{\beta} H^{2p-1}_{\mathrm{cont}}\big(\mathrm{GL}_n(\mathbb{C}), \mathbb{R}(p-1)\big). \end{aligned}$$

The isomorphism α is defined using invariant forms in the homogeneous space $U_n \cong U_n \backslash (U_n \oplus U_n)$. The isomorphism Id is obtained by identifying both spaces with the same subspace of $H^{2p-1}(\mathfrak{gl}_n(\mathbb{C}), \mathfrak{u}_n, \mathbb{C})$. And β is the inverse of the van Est isomorphism. In these groups $\mathfrak{gl}_n(\mathbb{C})$ is viewed as a real Lie algebra.

We will also denote by Bo_p the corresponding element in the group $H^{2p-1}\big(\mathfrak{gl}_n(\mathbb{C}), \mathfrak{u}_n, \mathbb{R}(p-1)\big)$. Moreover, since $\mathfrak{u}_n$ is noncohomologous to zero in $\mathfrak{gl}_n(\mathbb{C})$ (see Definition 5.40) the morphism

$$H^{2p-1}\big(\mathfrak{gl}_n(\mathbb{C}), \mathfrak{u}_n, \mathbb{R}(p-1)\big) \to H^{2p-1}\big(\mathfrak{gl}_n(\mathbb{C}), \mathbb{R}(p-1)\big)$$

is injective. Therefore the element Bo_p is determined by its image in the group $H^{2p-1}\big(\mathfrak{gl}_n(\mathbb{C}), \mathbb{R}(p-1)\big)$. We will also denote this element by Bo_p.

9.7. Explicit Representatives of the Borel Element

Let us determine explicitly a representative of Bo_p in the group $H^{2p-1}\big(\mathfrak{gl}_n(\mathbb{C}), \mathfrak{u}_n, \mathbb{R}(p-1)\big)$. For explicit representatives of the Borel element in continuous cohomology, the reader is referred to [**23, 36**].

By Example 5.37, the class of $\mathfrak{s}(\mathrm{ch}_p)$ is represented in the group $H^{2p-1}\big(\mathfrak{u}_n, \mathbb{R}(p)\big)$, by the form Φ_{2p-1} given by

$$\Phi_{2p-1}\Big(\bigwedge_{j=1}^{2p-1} x_j\Big) = \frac{(-1)^{p-1}(p-1)!}{(2p-1)!} \sum_{\sigma\in\mathfrak{S}^{2p-1}} (-1)^{\sigma} \operatorname{Tr}(x_{\sigma(1)} \circ \cdots \circ x_{\sigma(2p-1)}).$$

The isomorphism $\alpha\colon H^{2p-1}\big(\mathrm{U}_n, \mathbb{R}(p)\big) \to H^{2p-1}\big(\mathfrak{u}_n \oplus \mathfrak{u}_n, \mathfrak{u}_n, \mathbb{R}(p)\big)$ is obtained viewing U_n as the homogeneous space $\mathrm{U}_n \setminus \mathrm{U}_n \times \mathrm{U}_n$, where the inclusion $\mathrm{U}_n \to \mathrm{U}_n \times \mathrm{U}_n$ is given by $M \mapsto (\overline{M}, M)$. Therefore the projection $\varphi\colon \mathrm{U}_n \times \mathrm{U}_n \to \mathrm{U}_n$ is given by $\varphi(M,N) \mapsto M^t N$. The induced map $\varphi_*\colon \mathfrak{u}_n \oplus \mathfrak{u}_n \to \mathfrak{u}_n$ is given by $(x,y) \mapsto x^t + y$. Therefore $\alpha\big(\mathfrak{s}(\mathrm{ch}_p)\big)$ is represented by the form ω given by

$$\omega\Big(\bigwedge_{j=1}^{2p-1} (x_j, y_j)\Big) = \Phi_{2p-1}\Big(\bigwedge_{j=1}^{2p-1} x_j^t + y_j\Big).$$

This form extends to a $\mathbb{C}$-linear form

$$\omega_{\mathbb{C}} \in E^{2p-1}(\mathfrak{gl}_n(\mathbb{C}) \oplus \mathfrak{gl}_n(\mathbb{C}), \mathbb{C}) = E^{2p-1}(\mathfrak{gl}_n(\mathbb{C}) \otimes_{\mathbb{R}} \mathbb{C}, \mathbb{C})$$

given by the same formula. The inclusion

$$\mathfrak{gl}_n(\mathbb{C}) \to \mathfrak{gl}_n(\mathbb{C}) \otimes \mathbb{C} \to \mathfrak{gl}_n(\mathbb{C}) \oplus \mathfrak{gl}_n(\mathbb{C})$$

is given by $x \mapsto (\bar{x}, x)$. Thus we obtain:

PROPOSITION 9.25. *The Borel element* Bo_p *is represented by the form*

$$\bigwedge_{j=1}^{2p-1} x_j \mapsto \Phi_{2p-1}\Big(\bigwedge_{j=1}^{2p-1} \bar{x}_j^t + x_j\Big),$$

which is an element of $E^{2p-1}\big(\mathfrak{gl}_n(\mathbb{C}), \mathfrak{u}_n, \mathbb{R}(p-1)\big)$.

The image of the representative we have obtained in the group $E^{2p-1}\big(\mathfrak{gl}_n(\mathbb{C}), \mathbb{R}(p-1)\big)$ is not invariant under the action θ of $\mathfrak{gl}_n(\mathbb{C})$. For comparison with Beilinson's regulator we will need a representative of Bo_p in the group $E^{2p-1}\big(\mathfrak{gl}_n(\mathbb{C}), \mathbb{R}(p-1)\big)_{\theta=0}$. To obtain it we will use that $\mathfrak{s}(\mathrm{ch}_p)$ is primitive. Let us denote by μ the product in the group U_n:

$$\mu\colon \mathrm{U}_n \times \mathrm{U}_n \to \mathrm{U}_n$$
$$(M, N) \mapsto MN.$$

The fact that $\mathfrak{s}(\mathrm{ch}_p)$ is primitive implies that

$$\mu^*\big(\mathfrak{s}(\mathrm{ch}_p)\big) = \mathfrak{s}(\mathrm{ch}_p) \otimes 1 + 1 \otimes \mathfrak{s}(\mathrm{ch}_p).$$

Since there is a commutative diagram

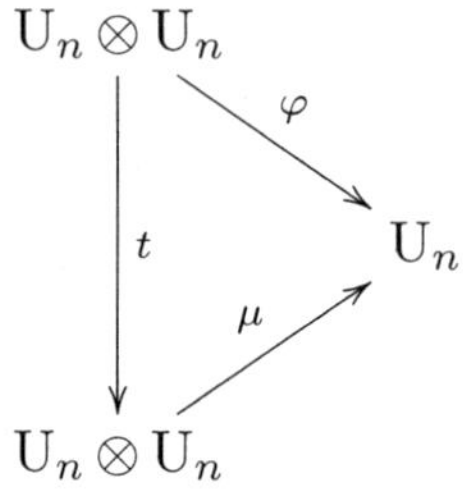

with $t(M, N) = (M^t, N)$ we obtain that $\alpha\big(\mathfrak{s}(\mathrm{ch}_p)\big)$ is represented, in the group $H^{2p-1}\big(\mathfrak{u}_n \oplus \mathfrak{u}_n, \mathbb{R}(p)\big)$ by the form

$$\omega'\Big(\bigwedge_{j=1}^{2p-1}(x_j, y_j)\Big) = \Phi_{2p-1}\Big(\bigwedge_{j=1}^{2p-1} x_j^t\Big) + \Phi_{2p-1}\Big(\bigwedge_{j=1}^{2p-1} y_j\Big).$$

Observe that this form is not an element of $E^{2p-1}\big(\mathfrak{u}_n \oplus \mathfrak{u}_n, \mathfrak{u}_n, \mathbb{R}(p)\big)$. Applying to ω' the same process applied to ω, we obtain that Bo_p in represented by the form

$$\bigwedge_{j=1}^{2p-1} x_j \mapsto \Phi_{2p-1}\Big(\bigwedge_{j=1}^{2p-1} \bar{x}_j^t\Big) + \Phi_{2p-1}\Big(\bigwedge_{j=1}^{2p-1} x_j^t\Big)$$

By the properties of the trace of a matrix is easy to see that

$$\Phi_{2p-1}\Big(\bigwedge_{j=1}^{2p-1} \bar{x}_j^t\Big) = (-1)^{p-1}\Phi_{2p-1}\Big(\bigwedge_{j=1}^{2p-1} x_j\Big)^{-}.$$

Let us denote by $\pi_{p-1}\colon \mathbb{C} \to R(p-1)$ the projection given by

$$\pi_{p-1}(x) = \frac{1}{2}(x + (-1)^{p-1}\overline{x}).$$

Then

PROPOSITION 9.26. *The image of the Borel element* Bo_p *in the group* $H^{2p-1}\big(\mathfrak{gl}_n(\mathbb{C}), \mathbb{R}(p-1)\big)$ *is represented by the form* $2\pi_{p-1} \circ \Phi_{2p-1}$.

CHAPTER 10

Beilinson's Regulator

10.1. Deligne–Beilinson Cohomology

In this section we will recall the definition and some properties of Deligne-Beilinson cohomology. For more details the reader is referred to [**1, 2, 27**].

Let $\overline{X}$ be a proper smooth algebraic variety over $\mathbb{C}$. Let $D \subset \overline{X}$ be a normal crossings divisor and let $X = \overline{X} - D$. We will consider X and $\overline{X}$ as complex analytical manifolds. Then Ω^*_X will denote the sheaf of holomorphic forms on X. Let us denote by $j\colon X \to \overline{X}$ the inclusion. Let $z_1 \cdots z_k = 0$ be a local equation for D. Then the sheaf of holomorphic forms on X with logarithmic poles along D, denoted $\Omega^*_{\overline{X}}(\log D)$, is the subalgebra of $j_*\Omega^*_X$ generated locally by $\Omega^*_{\overline{X}}$ and the sections

$$\frac{dz_i}{z_i}, \quad \text{for } i = 1, \ldots, k.$$

Since j is affine $Rj_*\Omega^*_X = j_*\Omega^*_X$. In the derived category of sheaves over $\overline{X}$ there are quasi-isomorphisms

$$Rj_*\mathbb{C} \to Rj_*\Omega^*_X = j_*\Omega^*_X \leftarrow \Omega^*_{\overline{X}}(\log D). \tag{10.1}$$

On the complex $\Omega^*_{\overline{X}}(\log D)$ there is defined a Hodge filtration

$$F^p\Omega^*_{\overline{X}}(\log D) = \bigoplus_{p' \geq p} \Omega^{p'}_{\overline{X}}(\log D).$$

In view of the quasi-isomorphisms (10.1), this filtration provides the cohomology of X, $H^*(X, \mathbb{C})$ with a Hodge filtration. This is the Hodge filtration of the mixed Hodge structure of $H^*(X, \mathbb{C})$ [**19**].

Let Λ be a subring of $\mathbb{C}$ and let us denote by $\Lambda(p)$ the subgroup $(2\pi i)^p\Lambda \subset \mathbb{C}$. Let us consider the complex of sheaves on $\overline{X}$

$$\Lambda(p)_{\mathcal{D}} = s(Rj_*\Lambda(p) \oplus F^p\Omega^*_{\overline{X}}(\log D) \xrightarrow{u} j_*\Omega^*_X), \tag{10.2}$$

where $s(\ )$ means the simple complex of a morphism of complexes (i.e., the cône shifted by one) and $u(r, f) = r - f$. In other words,

$$\Lambda(p)^n_{\mathcal{D}} = Rj_*\Lambda(p)^n \oplus F^p\Omega^n_{\overline{X}}(\log D) \oplus j_*\Omega^{n-1}_X,$$

with

$$d(r, f, \omega) = (dr, df, r - f - d\omega). \tag{10.3}$$

DEFINITION 10.1. The *Deligne–Beilinson cohomology* groups of X are the hypercohomology of the complex of sheaves $\Lambda(p)_{\mathcal{D}}$. These cohomology groups will be denoted by $H^m_{\mathcal{D}}(X,\Lambda(p))$.

There are obvious morphisms of complexes

$$u_1\colon \Lambda(p)_{\mathcal{D}} \to Rj_*\Lambda(p),$$
$$u_2\colon \Lambda(p)_{\mathcal{D}} \to F^p\Omega^*(\log D),$$

that induce morphisms

$$u_1^*\colon H^*_{\mathcal{D}}(X,\Lambda(p)) \to H^*(X,\Lambda(p)),$$
$$u_2^*\colon H^*_{\mathcal{D}}(X,\Lambda(p)) \to F^pH^*(X,\mathbb{C}).$$

LEMMA 10.2. *The diagram*

$$\begin{array}{ccc} H^*_{\mathcal{D}}(X,\Lambda(p)) & \xrightarrow{u_1^*} & H^*(X,\Lambda(p)) \\ {\scriptstyle u_2^*}\downarrow & & \downarrow \\ F^pH^*(X,\mathbb{C}) & \longrightarrow & H^*(X,\mathbb{C}) \end{array}$$

is commutative.

PROOF. Let $(r,f,\omega)\in\Lambda(p)^m_{\mathcal{D}}$ be a closed representative of a class $x\in H^m_{\mathcal{D}}(X,\Lambda(p))$. By (10.3), r and f are closed. Moreover r represents $u_1^*(x)$ and f represents $u_2(x)$. But, since $d\omega = r-f$, both elements represent the same class in $H^m(X,\mathbb{C})$. □

Alternatively we may represent Deligne–Beilinson cohomology by the complexes

$$\Lambda(p)_{\mathcal{D}} \cong s\bigl(Rj_*\Lambda(p) \to \text{cône}(F^p\Omega^*_{\overline{X}}(\log D) \to j_*\Omega^*_X)\bigr), \tag{10.4}$$

$$\Lambda(p)_{\mathcal{D}} \cong s\bigl(F^p\Omega^*_{\overline{X}}(\log D) \to \text{cône}(Rj_*\Lambda(p) \to j_*\Omega^*_X)\bigr). \tag{10.5}$$

From the presentations (10.2), (10.4) and (10.5), and the fact that the spectral sequence associated with the Hodge filtration degenerates in the term E_1, we obtain the exact sequences

$$H^{q-1}(X,\mathbb{C}) \to H^q_{\mathcal{D}}(X,\Lambda(p)) \to H^q(X,\Lambda(p)) \oplus F^pH^q(X,\mathbb{C}) \to \cdots, \tag{10.6}$$

$$\frac{H^{q-1}(X,\mathbb{C})}{F^pH^{q-1}(X,\mathbb{C})} \to H^q_{\mathcal{D}}(X,\Lambda(p)) \to H^q(X,\Lambda(p)) \to \cdots, \tag{10.7}$$

$$\frac{H^{q-1}(X,\mathbb{C})}{H^{q-1}(X,\Lambda(p))} \to H^q_{\mathcal{D}}(X,\Lambda(p)) \to F^pH^q(X,\mathbb{C}) \to \cdots. \tag{10.8}$$

For instance, if $\Lambda=\mathbb{Z}$ and $q=2p$ we obtain a short exact sequence

$$0 \to \frac{H^{2p-1}(X,\mathbb{C})}{F^pH^{2p-1}(X,\mathbb{C}) + H^{2p-1}(X,\mathbb{Z}(p))} \to H^{2p}_{\mathcal{D}}(X,\mathbb{Z}(p)) \to H^{2p}(X,\mathbb{Z}(p)) \cap F^pH^{2p}(X,\mathbb{C}) \to 0$$

Therefore, if X is projective, the group $H^{2p}_{\mathcal{D}}(X,\mathbb{Z}(p))$ is an extension between the intermediate Jacobian, $J_p(X)$, and the group of Hodge classes $H^{p,p}(X,\mathbb{Z}(p))$.

We will be interested in the real Deligne–Beilinson cohomology groups of X, $H^q_{\mathcal{D}}(X,\mathbb{R}(p))$. These groups can be represented using smooth differential forms. Let $\mathcal{E}^*$ be the complex of sheaves of $\mathbb{C}$-valued smooth differential forms on X, $\mathcal{E}^*_{\mathbb{R}}$ be the subcomplex of real valued forms and $\mathcal{E}^*_{\overline{X}}(\log D)$ be the sheaf of smooth differential forms with logarithmic singularities along D ([**14**]). We will denote by $E^*(X)$, $E^*_{\mathbb{R}}(X)$ and $E^*_{\overline{X}}(\log D)$ the corresponding complexes of global sections. If the compactification $\overline{X}$ and the divisor D are fixed, for simplicity, we will write $E^*_{\log}(X)=E^*_{\overline{X}}(\log D)$. Let us denote $E^*_{\mathbb{R}}(X,p) = (2\pi i)^p E^*_{\mathbb{R}}(X)$. Since $\mathcal{E}^*_{\mathbb{R}}$ is a resolution of $\mathbb{R}$, $\mathcal{E}^*_{\overline{X}}(\log D)$ is a resolution of $\Omega^*_{\overline{X}}(\log D)$ and $\mathcal{E}^*$ is a resolution of Ω^*, then *real Deligne–Beilinson cohomology* is the cohomology of the complex

$$D^*(X,p) = s\big(E^*_{\mathbb{R}}(X,p)\oplus F^pE^*_{\log}(X)\xrightarrow{u} E^*(X)\big).$$

Let us denote by $\pi_p\colon \mathbb{C}\to\mathbb{R}(p)$ the projection

$$\pi_p(x) = (x+(-1)^p\overline{x})/2.$$

In analogy with (10.5), the complex $D^*(X,p)$ is quasi-isomorphic to the complex

$$D^*_0(X,p) = s\big(F^pE^*_{\log}(X)\xrightarrow{\pi_{p-1}} E^*_{\mathbb{R}}(X,p-1)\big).$$

The definition of Deligne–Beilinson cohomology can be extended to real varieties. A smooth real variety $X_{\mathbb{R}}$, is a pair (X,σ), where X is a smooth complex variety and σ is an antilinear involution. Thus if $X_{\mathbb{R}}$ is a smooth real variety, we can define an involution F on $H^q_{\mathcal{D}}(X,\mathbb{R}(p))$, acting as σ on the space and as complex conjugation on the coefficients.

DEFINITION 10.3. Let $X_{\mathbb{R}}$ be a smooth real variety. Then the *real Deligne–Beilinson cohomology* groups of $X_{\mathbb{R}}$ are

$$H^q_{\mathcal{D}}(X_{\mathbb{R}},\mathbb{R}(p)) = H^q_{\mathcal{D}}(X,\mathbb{R}(p))^F,$$

the subspace of invariant elements under F.

EXAMPLE 10.4. Let us consider the variety $X=\operatorname{Spec}(\mathbb{C})$. From the exact sequence (10.7) it is clear that

$$H^n_{\mathcal{D}}(X,\mathbb{R}(p)) = \begin{cases}\mathbb{R}, & \text{if } n=0,\ p=0,\\ \mathbb{R}(p-1), & \text{if } n=1,\ p>0,\\ 0, & \text{otherwise.}\end{cases} \tag{10.9}$$

EXAMPLE 10.5. Let k be a number field. Let $X=\operatorname{Spec}(k)\otimes\mathbb{C}$ and let $X_{\mathbb{R}}$ be its natural structure of real variety. Then, for $p>0$,

$$H^1_{\mathcal{D}}(X_{\mathbb{R}},\mathbb{R}(p)) = \Big(\prod_{\Sigma}\mathbb{R}(p-1)\Big)^F.$$

REMARK 10.6. In general it is not obvious how to put an integral or even a rational structure on Deligne–Beilinson cohomology. Nevertheless, if X is defined over k, using the natural k-structures of Betti cohomology and of algebraic de Rham cohomology over k, one can define a k-structure on $H^n_{\mathcal{D}}(X, \mathbb{R}(p))$, for certain values of n and p. See [**2**] for the details. In the case of Example 10.5, since all the contributions to Deligne–Beilinson cohomology come from Betti cohomology, $H^1_{\mathcal{D}}(X_{\mathbb{R}}, \mathbb{R}(p))$ has a natural integral structure given by $\left(\prod_{\Sigma} \mathbb{Z}(p-1)\right)^F$.

10.2. Deligne–Beilinson Cohomology of $B.\mathrm{GL}_n(\mathbb{C})$

In this section we will compute the groups $H^{2p}_{\mathcal{D}}(B.\mathrm{GL}_n(\mathbb{C}), \mathbb{R}(p))$ and relate them with the continuous cohomology groups of $\mathrm{GL}_n(\mathbb{C})$.

If $X.$ is a simplicial smooth complex variety then $D^*(X., p)$ is a cosimplicial complex and we may define

$$H^*_{\mathcal{D}}(X., \mathbb{R}(p)) = H^*\left(s\mathcal{N}(D(X., p))\right).$$

In particular, the exact sequences (10.6), (10.7) and (10.8) are also valid for simplicial varieties.

Since the group $H^{2p}(B.\mathrm{GL}_n(\mathbb{C}), \mathbb{C})$ is pure of type (p,p) (see [**20**]), then

$$\frac{H^{2p}(B.\mathrm{GL}_n(\mathbb{C}), \mathbb{C})}{F^p H^{2p}(B.\mathrm{GL}_n(\mathbb{C}), \mathbb{C})} = 0.$$

Moreover $H^{2p-1}(B.\mathrm{GL}_n(\mathbb{C}), \mathbb{C}) = 0$. Therefore by the exact sequence (10.7), the morphism

$$u_1^* \colon H^{2p}_{\mathcal{D}}(B.\mathrm{GL}_n(\mathbb{C}), \mathbb{R}(p)) \to H^{2p}(B.\mathrm{GL}_n(\mathbb{C}), \mathbb{R}(p)). \tag{10.10}$$

is an isomorphism.

From the stability of the de Rham cohomology of $B.\mathrm{GL}_n(\mathbb{C})$ we obtain also a stability theorem for the Deligne–Beilinson cohomology of $B.\mathrm{GL}_n(\mathbb{C})$. We will write

$$H^m_{\mathcal{D}}(B.\mathrm{GL}(\mathbb{C}), \mathbb{R}(p)) = \varprojlim_n H^m_{\mathcal{D}}(B.\mathrm{GL}_n(\mathbb{C}), \mathbb{R}(p)).$$

If n, m are in the stable range, then, for all p, the natural morphisms

$$H^m_{\mathcal{D}}(B.\mathrm{GL}(\mathbb{C}), \mathbb{R}(p)) \to H^m_{\mathcal{D}}(B.\mathrm{GL}_n(\mathbb{C}), \mathbb{R}(p))$$

are isomorphisms.

For any complex manifold X, and for $p \geq 1$, let us define a morphism of complexes

$$\eta \colon D^*(X, p) \to E^0_{\mathbb{R}}(X, p-1)[-1],$$

given, for $(r, f, \omega) \in D^j(X, p)$, by

$$\eta(r, f, \omega) = \begin{cases} \pi_{p-1}(\omega), & \text{if } j = 1, \\ 0, & \text{if } j \neq 1. \end{cases}$$

Here we consider $E^0_{\mathbb{R}}(X,\mathbb{R}(p-1))[-1]$ as a complex concentrated in degree one. In particular, if $X = \operatorname{Spec}\mathbb{C}$, this morphism induces the isomorphism

$$H^1_{\mathcal{D}}(X,\mathbb{R}(p)) \to \mathbb{R}(p-1).$$

For the classifying space we obtain a morphism of complexes

$$\mathcal{N}D^*(B.\mathrm{GL}_n(\mathbb{C}),p) \to \mathcal{N}E^0_{\mathbb{R}}(B.\mathrm{GL}_n(\mathbb{C}),p-1)[-1],$$

which induces a map

$$\eta^*\colon H^q_{\mathcal{D}}(B.\mathrm{GL}_n(\mathbb{C}),\mathbb{R}(p)) \to H^{q-1}_{\mathrm{cont}}(\mathrm{GL}_n(\mathbb{C}),\mathbb{R}(p-1)).$$

10.3. The Definition of Beilinson's Regulator

In Section 10.1 we have defined Deligne–Beilinson cohomology as the hypercohomology of certain complex of sheaves on the analytic topology of a compactification $\overline{X}$ of X. Nevertheless Deligne–Beilinson cohomology can also be defined as the hypercohomology of certain complex of sheaves over the Zariski topology of X (see [**2, 27**]). Therefore we can apply Gillet's construction of characteristic classes for higher K-theory [**29**]. Beilinson's regulator map is the Chern character map between higher K theory and real Deligne–Beilinson cohomology.

For simplicity, instead of giving the general definition of Beilinson's regulator map, we will give an ad hoc definition in the case of number fields. The reader is referred to [**2**] for the general definition. Note that the definition presented here is simply a down to earth version of the definition in [**56**] (see also [**57**]).

Let us fix an integer p and let n be such that $n, 2p$ are in the stable range. Let us denote by $B.\mathrm{GL}_n(\mathbb{C})_s$ the simplicial set underlying $B.\mathrm{GL}_n(\mathbb{C})$. We can consider $\operatorname{Spec}(\mathbb{C})$ as a complex manifold with only one point. Let us write

$$B.\mathrm{GL}_n(\mathbb{C})^\delta = \operatorname{Spec}(\mathbb{C}) \times B.\mathrm{GL}_n(\mathbb{C})_s.$$

There is a natural morphism of simplicial complex manifolds

$$\mathrm{ev}\colon B.\mathrm{GL}_n(\mathbb{C})^\delta \to B.\mathrm{GL}_n(\mathbb{C}),$$

that induces a morphism

$$\mathrm{ev}^*\colon H^{2p}_{\mathcal{D}}(B.\mathrm{GL}_n(\mathbb{C}),\mathbb{R}(p)) \to H^{2p}_{\mathcal{D}}(B.\mathrm{GL}_n(\mathbb{C})^\delta,\mathbb{R}(p)).$$

Using equation (10.9), and Proposition 2.18, we can compute the Deligne–Beilinson cohomology of $B.\mathrm{GL}_n(\mathbb{C})^\delta$ and we obtain, for $p \geq 1$,

$$\begin{aligned} H^{2p}_{\mathcal{D}}(B.\mathrm{GL}_n(\mathbb{C})^\delta,\mathbb{R}(p)) &= H^{2p-1}\Big(B.\mathrm{GL}_n(\mathbb{C})^\delta, H^1_{\mathcal{D}}(\operatorname{Spec}(\mathbb{C}),\mathbb{R}(p))\Big) \\ &= \operatorname{Hom}\big(H_{2p-1}(B.\mathrm{GL}_n(\mathbb{C})^\delta,\mathbb{R}),\mathbb{R}(p-1)\big) \\ &= \operatorname{Hom}\big(H^{\mathrm{group}}_{2p-1}(\mathrm{GL}_n(\mathbb{C}),\mathbb{R}),\mathbb{R}(p-1)\big). \end{aligned}$$

By the stability of group homology (Theorem 6.5) we obtain a stability result for $H^{2p}_{\mathcal{D}}\big(B.\mathrm{GL}_n(\mathbb{C})^\delta, \mathbb{R}(p)\big)$. As usual we write

$$B.\mathrm{GL}(\mathbb{C})^\delta = \varinjlim B.\mathrm{GL}_n(\mathbb{C})^\delta,$$

and we define $H^{2p}_{\mathcal{D}}\big(B.\mathrm{GL}(\mathbb{C})^\delta, \mathbb{R}(p)\big)$ accordingly. The map ev^* induces a map

$$\mathrm{ev}^*\colon H^{2p}_{\mathcal{D}}\big(B.\mathrm{GL}(\mathbb{C}), \mathbb{R}(p)\big) \to H^{2p}_{\mathcal{D}}\big(B.\mathrm{GL}(\mathbb{C})^\delta, \mathbb{R}(p)\big).$$

Therefore, to each class $\alpha \in H^{2p}_{\mathcal{D}}\big(B.\mathrm{GL}(\mathbb{C}), \mathbb{R}(p)\big)$, we can associate a morphism

$$H_{2p-1}(B.\mathrm{GL}(\mathbb{C})^\delta, \mathbb{R}) \xrightarrow{\mathrm{ev}^*(\alpha)} \mathbb{R}(p-1).$$

If $\mathrm{ch}_p \in H^{2p}\big(B.\mathrm{GL}(\mathbb{C}), \mathbb{R}(p)\big)$ is the component of degree $2p$ of the twisted Chern character, we will also denote by ch_p its image in the group $H^{2p}_{\mathcal{D}}\big(B.\mathrm{GL}(\mathbb{C}), \mathbb{R}(p)\big)$ by the isomorphism $(u_1^*)^{-1}$ (see (10.10)).

DEFINITION 10.7. *Beilinson's regulator map*, for the field of complex numbers, is the composition

$$K_{2p-1}(\mathbb{C}) \xrightarrow{\mathrm{Hur}} H_{2p-1}(B.\mathrm{GL}(\mathbb{C})^\delta, \mathbb{R}) \xrightarrow{\mathrm{ev}^*(\mathrm{ch}_p)} \mathbb{R}(p-1).$$

This map will be denoted as $r_{\mathrm{Be},\mathbb{C}}$.

DEFINITION 10.8. Let k be a number field, $\mathfrak{o}$ its ring of integers and Σ the set of complex immersions of k. Then the *Beilinson's regulator map* for the ring $\mathfrak{o}$ is the composition

$$K_{2p-1}(\mathfrak{o}) \xrightarrow{\prod_\Sigma \sigma} \Big(\prod_\Sigma K_{2p-1}(\mathbb{C})\Big)^F \xrightarrow{\prod_\Sigma r_{\mathrm{Be},\mathbb{C}}} \Big(\prod_\Sigma \mathbb{R}(p-1)\Big)^F.$$

Let us see that Beilinson's regulator map is also determined by an element in continuous cohomology. Since $B.\mathrm{GL}_n(\mathbb{C})^\delta$ is discrete,

$$E^q(B.\mathrm{GL}_n(\mathbb{C})^\delta) = 0, \quad \text{for } q > 0,$$
$$E^0_{\mathbb{R}}(B.\mathrm{GL}_n(\mathbb{C})^\delta, p-1) = \mathcal{F}\big(B.\mathrm{GL}_n(\mathbb{C})^\delta, \mathbb{R}(p-1)\big),$$

where $\mathcal{F}\big(B_m\,\mathrm{GL}(\mathbb{C})^\delta, \mathbb{R}(p-1)\big)$ is the space of functions from $B_m\,\mathrm{GL}(\mathbb{C})^\delta$ to $\mathbb{R}(p-1)$. Therefore the morphisms ev^* is represented at the level of complexes by the morphism

$$\mathcal{N}D^j_0\big(B.\mathrm{GL}_n(\mathbb{C}), \mathbb{R}(p)\big) \to \mathcal{N}\mathcal{F}\big(B.\mathrm{GL}_n(\mathbb{C})^\delta, \mathbb{R}(p-1)\big)[-1],$$
$$(f,r) \mapsto \begin{cases} r, & \text{if } j=1,\\ 0, & \text{otherwise,}\end{cases}$$

Moreover, by the commutativity of the diagram

$$\begin{array}{ccc} H^{2p}_{\mathcal{D}}(B.\mathrm{GL}_n(\mathbb{C}),\mathbb{R}(p)) & \xrightarrow{\mathrm{ev}^*} & H^{2p}_{\mathcal{D}}(B.\mathrm{GL}_n(\mathbb{C})^{\delta},\mathbb{R}(p)) \\ {\scriptstyle \eta^*}\downarrow & & \downarrow \\ H^{2p-1}_{\mathrm{cont}}(\mathrm{GL}_n(\mathbb{C}),\mathbb{R}(p-1)) & \longrightarrow & H^{2p-1}_{\mathrm{group}}(\mathrm{GL}_n(\mathbb{C}),\mathbb{R}(p-1)) \end{array}$$

we obtain that Beilinson's regulator map is determined by the elements

$$\eta^*(\mathrm{ch}_p) \in H^{2p-1}_{\mathrm{cont}}\big(\mathrm{GL}_n(\mathbb{C}),\mathbb{R}(p-1)\big).$$

These elements will be called *Beilinson elements* and denoted by Be_p.

10.4. The Comparison Between the Regulators

In this section we will compare Beilinson's regulator, r_{Be}, and the renormalized Borel's regulator, r_{Bo}. Observe that, in view of Proposition 9.21 and Definition 10.8 it is enough to treat the case of $\mathrm{Spec}(\mathbb{C})$. Clearly, to compare the morphisms we can compare the elements Bo_p and Be_p in the group $H^{2p-1}_{\mathrm{cont}}\big(\mathrm{GL}_n(\mathbb{C}),\mathbb{R}(p-1)\big)$, for $n, 2p$ in the stable range (see Definition 9.4).

THEOREM 10.9.

$$\mathrm{Bo}_p = 2\mathrm{Be}_p.$$

PROOF. Since we have an explicit representative of Bo_p in the group $E^{2p-1}\big(\mathfrak{gl}_n(\mathbb{C}),\mathbb{R}(p-1)\big)$, the strategy will be to look for an explicit representative of Be_p in the same group.

Let us choose a compactification $\overline{B.\mathrm{GL}_n(\mathbb{C})}$ of $B.\mathrm{GL}_n(\mathbb{C})$ with the divisor at infinity $D.$ a simplicial normal crossing divisor. Let

$$j\colon\ B.\mathrm{GL}_n(\mathbb{C}) \to \overline{B.\mathrm{GL}_n(\mathbb{C})}$$

be the inclusion. Let us denote by $\Omega^*_{\log}$ the sheaf of holomorphic forms on $\overline{B.\mathrm{GL}_n(\mathbb{C})}$ with logarithmic poles along $D.$. Let $\mathcal{E}^0_{\mathbb{R}}$ be the sheaf of real C^∞ functions on $B.\mathrm{GL}_n(\mathbb{C})$ and $\mathcal{E}^0_{\mathbb{R}}(p) = (2\pi i)^p\mathcal{E}^0_{\mathbb{R}}$.

Let $\mathcal{I} \subset \mathcal{O}_{B_1\mathrm{GL}_n(\mathbb{C})} = \mathcal{O}_{\mathrm{GL}_n(\mathbb{C})}$ be the ideal $\operatorname{Ker}\sigma_0^*$. It is the ideal of the point e. Let $\mathcal{J}$ be the ideal of $\mathcal{O}_{B.\mathrm{GL}_n}(\mathbb{C})$ generated by $\mathcal{I}^2$. Let $B.^{(1)}\mathrm{GL}_n(\mathbb{C})$ be the subscheme defined by the ideal $\mathcal{J}$. We will consider it with the analytical topology. Observe that $B.^{(1)}\mathrm{GL}_n(\mathbb{C})$ is concentrated on one point in each degree. Let us denote by

$$i\colon\ B.^{(1)}\mathrm{GL}_n(\mathbb{C}) \to B.\mathrm{GL}_n(\mathbb{C})$$

the inclusion.

For $p \geq 1$, we have the following commutative diagram of sheaves

$$\begin{array}{ccccc}
F^p\Omega^*_{\log} & \xleftarrow{u_2^*} & s(Rj_*\mathbb{R}(p) \oplus F^p\Omega^*_{\log} \to j_*\Omega^*) & \xrightarrow{\eta} & j_*\mathcal{E}^0_{\mathbb{R}}(p-1)[-1] \\
\downarrow & & \downarrow & & \downarrow \\
i^*F^p\Omega^*_{\log} & \longleftarrow & i^*s(Rj_*\mathbb{R}(p) \oplus F^p\Omega^*_{\log} \to j_*\Omega^*) & \xrightarrow{\eta} & i^*\mathcal{E}^0_{\mathbb{R}}(p-1)[-1] \\
& & \downarrow & \nearrow_{\eta'} & \\
& & i^*(Rj_*\mathbb{R}(p) \to j_*\mathcal{O}) & &
\end{array}$$

where the morphism η' is induced by η. The fact that $B^{(1)}_{\cdot}\mathrm{GL}_n(\mathbb{C})$ has support on only one point has several consequences. First, the sheaves of the second and third row are, in fact, cosimplicial abelian groups. Moreover $i^*\Omega^*_{\log} = i^*\Omega^* = \Omega^*(B^{(1)}_{\cdot}G)$. Finally $i^*(Rj_*\mathbb{R}) = \mathbb{R}$.

Let us compute the normalization of the cosimplicial abelian groups of the second and third row.

$$\mathcal{N}\big(i^*(Rj_*\mathbb{R})\big) = \mathcal{N}(\mathbb{R}) = \mathbb{R},$$

a complex concentrated in degree zero. Therefore for computing the cohomology in degree greater than zero we can ignore it. By Proposition 8.9,

$$\mathcal{N}i^*\mathcal{O} = E^*(\mathfrak{gl}_n(\mathbb{C}), \mathbb{C}).$$

By Proposition 8.10,

$$\mathcal{N}i^*\Omega^* = W^{*,\cdot}\big(\mathfrak{gl}_n(\mathbb{C})\big).$$

Hence we also have

$$\mathcal{N}i^*F^p\Omega^*_{\log} = F^pW^{*,\cdot}\big(\mathfrak{gl}_n(\mathbb{C})\big).$$

Note that, in the above equalities, $\mathfrak{gl}_n(\mathbb{C})$ is considered a complex Lie algebra. Finally, by Proposition 8.14

$$\mathcal{N}i^*\mathcal{E}^0_{\mathbb{R}}(p-1) = E^*\big(\mathfrak{gl}_n(\mathbb{C}), \mathbb{R}(p-1)\big).$$

In the last equality, $\mathfrak{gl}_n(\mathbb{C})$ is viewed as a real Lie algebra. Therefore, taking cohomology, we obtain the diagram

$$\begin{array}{ccccc}
H^{2p}(B.G, \mathbb{C}) & \xleftarrow{(1)} & H^{2p}_{\mathcal{D}}\big(B.G, \mathbb{R}(p)\big) & \xrightarrow{\eta^*} & H^{2p-1}_{\mathrm{cont}}\big(G, \mathbb{R}(p-1)\big) \\
\downarrow & & \downarrow & & \downarrow{\scriptstyle (3)} \\
H^{2p}(F^pW) & \xleftarrow{\cong} & H^{2p}\big(s(F^pW \to W)\big) & \longrightarrow & H^{2p-1}\big(\mathfrak{g}^{\mathbb{R}}, \mathbb{R}(p-1)\big) \\
& & \downarrow & \nearrow_{\pi_{p-1}} & \\
& & H^{2p-1}(\mathfrak{g}, \mathbb{C}) & &
\end{array}$$

(with an arrow (2) from $H^{2p}(B.G, \mathbb{C})$ to $H^{2p-1}(\mathfrak{g}, \mathbb{C})$)

where we have written G for $\mathrm{GL}_n(\mathbb{C})$, $\mathfrak{g}$ for the complex Lie algebra $\mathfrak{gl}_n(\mathbb{C})$, $\mathfrak{g}^{\mathbb{R}}$ for the Lie algebra $\mathfrak{gl}_n(\mathbb{C})$ viewed as a real Lie algebra, and W for $W\big(\mathfrak{gl}_n(\mathbb{C})\big)$.

By Lemma 10.2, the morphism (1) is the isomorphism u_1^* (10.10), followed by the inclusion

$$H^{2p}\big(B.\mathrm{GL}_n(\mathbb{C}),\mathbb{R}(p)\big) \to H^{2p}(B.\mathrm{GL}_n(\mathbb{C}),\mathbb{C}).$$

By Theorem 8.12 and Proposition 5.33, the morphism (2) is the suspension followed by the identification

$$H^{2p-1}(\mathrm{GL}_n(\mathbb{C}),\mathbb{C}) = H^{2p-1}(\mathfrak{gl}_n(\mathbb{C}),\mathbb{C}).$$

By Theorem 8.15 the morphism (3) is the van Est isomorphism followed by the inclusion

$$H^{2m-1}\big(\mathfrak{gl}_n(\mathbb{C}),\mathfrak{u}_n,\mathbb{R}(m-1)\big) \to H^{2m-1}\big(\mathfrak{gl}_n(\mathbb{C}),\mathbb{R}(m-1)\big).$$

Therefore we have proved the following result

LEMMA 10.10. *The image of the Beilinson element under the composition*

$$\begin{aligned} H^{2p-1}_{\mathrm{cont}}\big(\mathrm{GL}_n(\mathbb{C}),\mathbb{R}(p-1)\big) &\xrightarrow{\mathrm{vEst}} H^{2p-1}\big(\mathfrak{gl}_n(\mathbb{C}),\mathfrak{u}_n,\mathbb{R}(p-1)\big) \\ &\longrightarrow H^{2p-1}\big(\mathfrak{gl}_n(\mathbb{C}),\mathbb{R}(p-1)\big) \end{aligned}$$

agrees with the image of ch_p *under the composition*

$$\begin{aligned} H^{2p}\big(B.\mathrm{GL}_n(\mathbb{C}),\mathbb{R}(p)\big) &\longrightarrow H^{2p}(B.\mathrm{GL}_n(\mathbb{C}),\mathbb{C}) \\ &\xrightarrow{\mathfrak{s}} H^{2p-1}(\mathrm{GL}_n(\mathbb{C}),\mathbb{C}) \\ &=\!=\!= H^{2p-1}(\mathfrak{gl}_n(\mathbb{C}),\mathbb{C}) \\ &\xrightarrow{\pi_{p-1}} H^{2p-1}\big(\mathfrak{gl}_n(\mathbb{C}),\mathbb{R}(p-1)\big). \end{aligned}$$

Let Φ_{2p-1} be the forms used in Section 9.7 to obtain an explicit representative of Borel's element. Then, by Proposition 5.34, Example 5.37 and Lemma 10.10 we obtain:

PROPOSITION 10.11. *The image of* Be_p *in* $H^{2p-1}\big(\mathfrak{gl}_n(\mathbb{C}),\mathbb{R}(p-1)\big)$ *is represented by the form* $\pi_{p-1}\circ\Phi_{2p-1}$.

This proposition and Proposition 9.26 conclude the proof of the theorem. □

Bibliography

1. A. A. Beilinson, *Notes on absolute Hodge cohomology*, Applications of Algebraic K-Theory to Algebraic Geometry and Number Theory (S. J. Bloch, ed.), Contemp. Math., vol. 55, Amer. Math. Soc., Providence, RI, 1983, pp. 35–68.
2. ———, *Higher regulators and values of L-functions*, J. Soviet Math. **30** (1985), 2036–2070.
3. S. J. Bloch, *Higher regulators, algebraic K-theory and zeta functions of elliptic curves*, CRM Monogr. Ser., vol. 11, Amer. Math. Soc., Providence, RI, 2000.
4. A. Borel, *Values of zeta-functions at integers, cohomology and polylogarithms*, Current Trends in Mathematics and Physics (S. D. Adhikari, ed.), Narosa, New Delhi, 1995, pp. 1–44.
5. ———, *Sur la cohomologie des espaces fibrés principaux et des espaces homogènes de groupes de Lie compacts*, Ann. of Math. (2) **57** (1953), 115–207.
6. ———, *Introduction aux groupes arithmétiques*, Actualités Sci. Indust., vol. 1341, Hermann, Paris, 1969.
7. ———, *Stable real cohomology of arithmetic groups*, Ann. Sci. École Norm. Sup. (4) **7** (1974), 235–272.
8. ———, *Cohomologie de SL_n et valeurs de fonctions zeta aux points entiers*, Ann. Scuola Norm. Sup. Pisa Cl. Sci. (4) **4** (1977), 613–636; Errata, **7**, (1980), 373.
9. A. Borel and N. Wallach, *Continuous cohomology, discrete subgroups, and representations of reductive groups*, Ann. of Math. Stud., vol. 94, Princeton Univ. Press, Princeton, NJ, 1980.
10. R. Bott, *The stable homotopy of the classical groups*, Ann. of Math. (2) **70** (1959), 313–337.
11. N. Bourbaki, *Éléments de mathématique.* XXVI. *Groupes et algèbres de Lie. Chapitre* I. *Algèbres de Lie*, Actualités Sci. Indust., vol. 1285, Hermann, Paris, 1960.
12. A. K. Bousfield and D. M. Kan, *Homotopy limits, completions and localizations*, Lecture Notes in Math., vol. 304, Springer-Verlag, Berlin-New York, 1972.
13. K. S. Brown and S. M. Gersten, *Algebraic K-theory as generalized sheaf cohomology*, Algebraic K-theory. I. Higher K-Theories (Seattle, WA, 1972) (H. Bass, ed.), Lecture Notes in Math., vol. 341, Springer-Verlag, Berlin-New York, 1973, pp. 266–292.
14. J. I. Burgos, *A C^∞ logarithmic Dolbeault complex*, Compositio Math. **92** (1994), 61–86.
15. H. Cartan, *Démonstration homologique des théorèmes de périodicité de Bott.* I, Exp. 16, Séminaire Henri Cartan 12ᵉ année: 1959/60, École Normale Supérieure, Secrétariat mathématique, Paris, 1961; II, Exp. 17; III, Exp. 18.
16. H. Cartan (ed.), Séminaire Henri Cartan 12ᵉ année: 1959/60, École Normale Supérieure, Secrétariat mathématique, Paris, 1961
17. ———, *La transgression dans un groupe de Lie et dans un espace fibré principal*, Œuvres, vol. III (R. Remmert and J-P. Serre, eds.), Springer-Verlag, Berlin-New York, 1979, pp. 1255–1267
18. ———, *Notions d'algèbre différentielle*, Œuvres, vol. III (R. Remmert and J-P. Serre, eds.), Springer-Verlag, Berlin-New York, 1979, pp. 1268–1282.

19. P. Deligne, *Théorie de Hodge*. II, Inst. Hautes Études Sci. Publ. Math. **40** (1972), 5–57.
20. ______, *Théorie de Hodge*. III, Inst. Hautes Études Sci. Publ. Math. **44** (1975), 5–77.
21. A. Dold, *Lectures on algebraic topology*, Grundlehren Math. Wiss., vol. 200, Springer-Verlag, New York-Berlin, 1972.
22. A. Douady, *Périodicité du groupe unitaire*, Exp. 11–12, Séminaire Henri Cartan 12e année: 1959/60, École Normale Supérieure, Secrétariat mathématique, Paris, 1961.
23. J. L. Dupont, *Simplicial de Rham cohomology and characteristic classes of flat bundles*, Topology **15** (1976), 233–245.
24. ______, *Curvature and characteristic classes*, Lecture Notes in Math., vol. 640, Springer-Verlag, Berlin-New York,, 1978.
25. J. L. Dupont, R. Hain, and S. Zucker, *Regulators and characteristic classes of flat bundles*, The Arithmetic and Geometry of Algebraic Cycles (Banff, AB, 1998) (B. B. Gordon, J. D. Lewis, S. Müller-Stach, S. Saito, and N. Yui, eds.), CRM Proc. Lecture Notes, vol. 24, Amer. Math. Soc., Providence, RI, 2000, pp. 47–92.
26. S. Eilenberg and S. MacLane, *On the groups* $H(\Pi, n)$. II. *Methods of computation*, Ann. of Math. (2) **60** (1954), 49–139.
27. H. Esnault and E. Viehweg, *Deligne–Beilinson cohomology*, Beilinson's Conjectures on Special Values of L-Functions (M. Rapoport, N. Schappacher, and P. Schneider, eds.), Perspect. Math., vol. 4, Academic Press, Boston, MA, 1988, pp. 43–91.
28. E. M. Friedlander, *Étale homotopy of simplicial schemes*, Ann of Math. Stud., vol. 104, Princeton Univ. Press, Princeton, NJ, 1982.
29. H. Gillet, *Riemann–Roch theorems for higher algebraic K-theory*, Adv. in Math. **40** (1981), 203–289.
30. ______, *Universal cycle classes*, Compositio Math. **49** (1983), 3–49.
31. P. G. Goerss and J. F. Jardine, *Simplicial homotopy theory*, Progr. Math., vol. 174, Birkhaüser Verlag, Basel, 1999.
32. W. Greub, S. Halperin, and R. Vanstone, *Connections, curvature and cohomology. II. Lie groups, principal bundles, and characteristic classes*, Pure Appl. Math., vol. 47-II, Academic Press, New York-London, 1973.
33. ______, *Connections, curvature and cohomology*. III. *Cohomology of principal bundles and homogeneous spaces* Pure Appl. Math., vol. 47-III, Academic Press, New York-London, 1976.
34. B. Gross, *On the values of Artin L-functions*, 1979 (unpublished manuscript).
35. A. Guichardet, *Cohomologie des groupes topologiques et des algèbres de Lie*, Textes Math., vol. 2, CEDIC, Paris, 1980.
36. N. Hamida, *Description explicite du régulateur de Borel*, C. R. Acad. Sci. Paris Sr. I Math. **330** (2000), 169–172.
37. R. Hartshorne, *Algebraic geometry*, Grad. Texts in Math., vol. 52, Springer-Verlag, New York-Heidelberg, 1977.
38. F. Hirzebruch, *Topological methods in algebraic geometry*, 3rd ed., 2nd corr. print ed., Grundlehren Math. Wiss., vol. 131, Springer-Verlag, New York, 1978.
39. G. Hochschild, *The structure of Lie groups*, Holden Day Inc., San Francisco-London-Amsterdam, 1965.
40. A. Huber and J. Wildeshaus, *Classical motivic polilogarithm according to Beilinson and Deligne*, Doc. Math. **3** (1998), 27–133 (electronic).
41. W. van der Kallen, *Homology stability for linear groups*, Invent. Math. **60** (1980), 269–295.
42. S. Lichtenbaum, *Values of zeta functions, étale cohomology, and algebraic K-theory*, Algebraic K-Theory. II. "Classical" Algebraic K-Theory and Connections with Arithmetic (Seattle, WA, 1972) (H. Bass, ed.), Lecture Notes in Math., vol. 342, Springer, Berlin 1973, pp. 489–501.

43. S. MacLane, *Homology*, Grundlehren Math. Wiss., vol. 114, Springer-Verlag, Berlin-Göttingen-Heidelberg, 1963.
44. J. P. May, *Simplicial objects in algebraic topology*, Van Nostrand Math. Stud., vol. 11, Van Nostrand, Princeton, NJ-Toronto, ON-London, 1967.
45. J. McCleary, *User's guide to spectral sequences*, Math. Lecture Ser., vol. 12, Publish or Perish, Inc., Wilmington, DE, 1985.
46. J. W. Milnor and J. J. Moore, *On the structure of Hopf algebras*, Ann. of Math (2) **81** (1965), 211–264.
47. J. W. Milnor and J. S. Stasheff, *Characteristic classes*, Ann. of Math. Stud., vol. 76, Princeton Univ. Press, Princeton, NJ, 1974.
48. M. Mimura, *Homotopy theory of Lie groups*, Handbook of Algebraic Topology (I. M. James, ed.), North-Holland, Amsterdam, 1995, pp. 951–991.
49. J. C. Moore, *Semi-simplicial complexes and Postnikov systems*, Symposium Internacional de Topología Algebraica, Universidad Nacional Autónoma de México, Mexico City, 1958 pp. 232–247.
50. ———, *Algèbre homologique et homologie des espaces classifiants*, Exp. 7, Séminaire Henri Cartan 12e année: 1959/60, École Normale Supérieure, Secrétariat mathématique, Paris, 1961.
51. ———, *Algèbres de Hopf universelles*, Exp. 10, Séminaire Henri Cartan 12e année: 1959/60, École Normale Supérieure, Secrétariat mathématique, Paris, 1961.
52. ———, *Espaces classifiants*, Exp. 5, Séminaire Henri Cartan 12e année: 1959/60, École Normale Supérieure, Secrétariat mathématique, Paris, 1961.
53. ———, *La suspension*, Exp. 6, Séminaire Henri Cartan 12e année: 1959/60, École Normale Supérieure, Secrétariat mathématique, Paris, 1961.
54. V. Navarro Aznar, *Sur la théorie de Hodge-Deligne*, Invent. math. **90** (1987), 11–76.
55. M. Rapoport, *Comparison of the regulators of Beilinson and of Borel*, Beilinson's Conjectures on Special Values of L-Functions (M. Rapoport, N. Schappacher, and P. Schneider, eds.), Perspect. Math., vol. 4, Academic Press, Boston, MA, 1988, pp. 169–192.
56. P. Schneider, *Introduction to the Beilinson conjectures*, Beilinson's Conjectures on Special Values of L-Functions (M. Rapoport, N. Schappacher, and P. Schneider, eds.), Perspect. Math., vol. 4, Academic Press, Boston, MA, 1988, pp. 1–35.
57. C. Soulé, *Régulateurs*, Séminaire Bourbaki, vol. 1984/85, Astérisque, vol. 133-134, Soc. Math. France, Paris, 1986, pp. 237–253.
58. V. Srinivas, *Algebraic K-theory*, Progr. Math., vol. 90, Birkhäuser, Boston, MA, 1991.
59. A. A. Suslin, *Stability in algebraic K-theory*, Algebraic K-Theory. I (Oberwolfach, 1980), Lecture Notes in Math., vol. 966, Springer, Berlin, 1982, pp. 304–333.
60. M. Zisman, *Cohomologie des variétés de Stiefel*, Exp. 3, Séminaire Henri Cartan 12e année: 1959/60, École Normale Supérieure, Secrétariat mathématique, Paris, 1961.

Index